MW01365512

Peace Was
Our
Profession

We Were Crewdogs III

Edited by
Tommy Towery

Should the decision be made to publish a future volume of stories such as this and you want to participate, then please contact the editor.

The Strategic Air Command Coin depicted on the front cover can be obtained by contacting:

Lori Hein
Cold War Coins
2701 S. Avondale Ct.
Sioux Falls, SD 57110
(605) 275-6275

Book Contact Info:
Tommy Towery
5709 Pecan Trace
Memphis, TN 38135
ttowery@memphis.edu

www.wewerecrewdogs.com

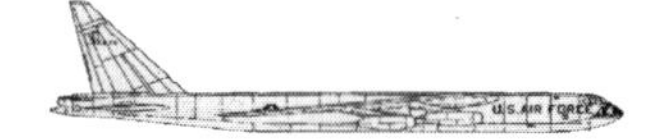

Table of Contents

Foreword

Chapter 1 - Military Careers

Chapter 2 - Cold War

Chapter 3 - Inflight Emergencies

Chapter 4 - Southeast Asia

Chapter 5 - Bar Stories

Chapter 6 - Lest We Forget

Résumés

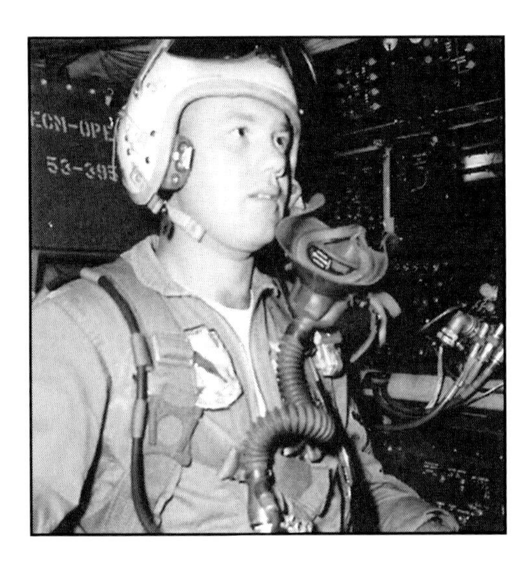

Dedication

This book is dedicated to George Donald Jackson, Major, USAF (Ret.), who was one of the contributing authors of not only this volume but also the previous one in this series. He epitomized the sacrifices many have made in their service to our country.

While aboard a routine B-52 training mission in 1961, his aircraft was shot down by the accidental firing of a Sidewinder missile from a New Mexico Air National Guard F-100 Super Sabre fighter. Three crewmembers were killed in that accident and as a result of his injuries George endured pain and suffering while he was hospitalized for an extended period of time. During his recovery he not only was grounded and lost his flight pay, but also lost a spot promotion that reduced his rank and subsequent income by one pay grade. He accepted these actions as few professionals would, and returned to flight status to continue to contribute his time and service to our freedom and make a career in the Air Force.

Early on George recognized the need to preserve such stories for his family and friends and worked through intense pain in his final days to continue to write down some of his favorite stories of his military career while he still could. In one of his final correspondences to me he wrote:

"It was a real joy to see the faces and hear the comments of family and friends as they read Volume II. I really appreciated your efforts that allowed me to express some of the experiences we all had during our BUFF years. They are written and published now and can't be taken back. The only thing I missed on the last one was not dedicating it to my recently departed wife of 46 years. She passed away In July 2004. These last three years have been much more difficult without her. Then on the other hand I'm glad she doesn't have to watch me go through this!"

George passed away on July 7, 2007, before he would ever see the final results of his last writings. Rest in peace, George – you helped preserve it.

- TT

Foreword

Peace [pees] - *noun* - The normal, nonwarring condition of a nation, group of nations, or the world.

Our [ou-er] - pro*noun* - Used to indicate a particular profession, nationality, political party, etc., that includes the speaker or writer.

Profession [pruh-fesh-uhn] - *noun* - An occupation that requires considerable training and specialized study.

Peace Was Our Profession – Foreword

Walter J. Boyne

The United States Air Force, like many organizations, is prey to the do-gooder who wants to make a name for him or herself by a facile idea, a phrase, or a concept that will, for the moment, attract attention, and then fade away without leaving a grace. Often they do so by creating mottos—"An Army of One" is a great example, but another one is not. "Peace is Our Profession" may have been created by some wordsmith, but it represented the heart and soul of the Strategic Air Command, for maintaining peace was SAC's profession, and not only that - being professional was essential to being a SAC crew-member.

It is startling to believe that (if my numbers are correct) some 90% of the people presently in the United States Air Force came in after SAC's untimely demise in 1992. I say untimely because the Air Force does not have the British sense of tradition, and does not retain designations with a proud heritage, but rather scrambles them into a new mix.

If ever a heritage should have been maintained it is that of the Strategic Air Command. The current Air Combat Command is a splendid outfit, but it would be more splendid, and more to the point, if it had retained the proud name of the Strategic Air Command. I know, as you do, that the change was made because the distinction between tactical and strategic bombing had become blurred, but it doesn't make any difference. The Air Combat Command has a strategic purpose, and it would be better served by having the name that General Curtis E.

LeMay made glorious through hard work and sacrifice: the Strategic Air Command.

I feel strongly about this because I have an attachment to aircraft and aviation that has dominated my life since I was four. In the intervening almost 75 years, I have devoted myself to aviation in a manner that today would be diagnosed as obsessive-compulsive and be treated with doses of Ritalin. And I have to tell you that the happiest of those days were spent in SAC, when I was a buck-green copilot flying B-50s and then B-47s. It is impossible to tell anyone else of the sheer pleasure of being the first crew-member to be at the airplane on some early morning, to crawl into the bays where the landing gear retracts, to check out the bomb bays, to be the first into the cockpit, reading the Form 1, talking to the crew chief, assessing the probability of taking off. It was even a pleasure to go to the mess hall and pick up the box lunches for the long flight, for it was a privilege to be in SAC, and to be able to fly in those glorious aircraft on those glorious missions.

One of the first things that came across to me as a green 2nd Lieutenant, fresh out of flying school, was that there was a vast pool of knowledge, service and good will in the enlisted men who made up the crews and who serviced the airplanes. I was astounded by how much they knew and how hard they worked. Their tolerance of my ignorance was equally astounding; later I realized they had seen a hundred second balloons come and go, and they knew that they really didn't matter. If they learned something and became competent, that was great; if they didn't, the enlisted crews would be there to save the day anyway.

And this vital human element was not confined to ground crews. Now there were some star turns in SAC, hot young officers who distinguished themselves, got spot promotions, and perhaps ultimately reached flag rank. But the real human element that distinguished SAC was the ordinary crew-member who did his job—aircraft commander, pilot, radar observer, navigator, or later, weapon system officer—and did it well, day in and day out, mission after mission. This was the heart of SAC, the combination of dedicated ground and flight crews who knew they were never going to get a spot promotion, who recognized that crew duties were a career dead end, but who also saw that SAC was the country's bulwark, and who were proud to serve under its banner.

Tommy Towery's books are a salute to the members of the great organization that was SAC, and a salute in the purest form, for they are generated by SAC personnel themselves. This insures that there is no (malarky is the polite term) in the writing, for the writers know that the readers know, and there is no way to fool them. It would be sad to think that SAC, which once had hundreds of bases around the world, with thousands of airplanes and hundreds of thousands of people, is no more. But that is not true. SAC exists still in the capability of Air Combat and Air Mobility Commands; it exists in the memory of everyone who belonged to SAC, and it exists because the United States is still a free nation. It would not be so, without the Strategic Air Command and its people.

Chapter One

Military Career [mil-i-ter-ee] [kuh-reer] *–noun* - An occupation or profession, especially one requiring special training, for, or pertaining to war - followed as one's lifework.

The author in 1964.

Crewdog Life on the Northern Tier

Bob Stewart

After arriving at Glasgow Air Force Base in east Montana following a long assignment on the Gulf Coast, I was told that only the most promising and outstanding personnel were assigned to the northern tier of SAC bases. They said that only the hale, hardy, and smart could withstand those frigid winters. I believed them. I was convinced that all the weaklings and dummies were down south in the warmer climes.

Years later while involved with Arc Light flights at Guam, we encountered other Crewdogs who had lived the life of Riley while basking at the sunny SAC bases. I discovered they were not weaklings and dummies at all. The Southerners told us Northern warriors that only the strongest and most brilliant were chosen to man those Southern paradises. I mean those that had Olympic-size swimming pools adjoining their alert facilities and 36-hole golf courses that rivaled Pinehurst Number One. It was those Southern Crewdogs with the dark sun tans and calloused golf hands that convinced us Eskimos that we indeed were the dumb, naïve and unfortunate souls!

It did not take me long to realize that Glasgow was isolated from the inhabited world. The Crewdogs assigned there were denied many pleasures of the civilized world. In other words we were not tempted to waste our spare time by golfing, swimming, camping and boating. So we entertained ourselves by diligently studying our Dash Ones, our tactical doctrine and our emergency war orders. At least that is what I think I remember. But we did on occasion attempt to make life enjoyable during our days off duty.

When my crew training was completed at Roswell, New Mexico, in December of 1962, I drove my family through minus 35° weather to Glasgow. We were pulling our home-made box trailer containing our living essentials. About 30 miles from Glasgow, a tire blew on the trailer. While I was digging out my winter clothing from the car trunk, a concerned rancher stopped by and offered help in changing the tire. I tried my best to convince him I had plenty of heavy clothing and could do it myself, but he did not believe me and practically forced me back inside the warm car while he did the changeover.

Our first day in the town of Glasgow, located 19 miles south of the base, was spent in a motel on the edge of town. During the night we heard an awful scraping racket. Then someone knocked on our motel door and exclaimed that thieves had tried to make off with our trailer! They would have succeeded except they neglected to fold up the wheel on the trailer tongue. It had dragged the blacktop when they were pulling it up from a drainage ditch. My family decided right then and there that they did not like this assignment!

A family of four Montana natives froze to death in their car during one blizzard on the road between the base and town. This occurred just prior to the opening of the new base. Because of that tragedy, everyone later was cautioned to carry blankets and candles and matches in their car during any winter traveling. And all were cautioned to forego travel when blizzard conditions were forecast.

One time my family decided to entertain our crew with something different. We decided to host a daytime ice skating party on a ranch lake near the base. It was early winter. My wife had made a huge pot of vegetable soup, enough to feed about 20 people. The covered pot was loaded into the trunk of my Chrysler Newport, and off we went. Soon we headed up a small hill and then crossed over a rough cattle guard. I guess I was driving too fast to make a smooth passage across the guard, as the lid bounced off the pot and most of the soup splashed all over the inside of the car trunk. We had frozen soup in that car for the next six months! We kept chipping away until the spring thaw which began in June.

During the long winter months Glasgow Crewdogs seldom entertained family in the visitor area near the alert facility gate. It was too difficult to keep a car warm to visit very long. And trying to kiss one's sweetie while wearing Arctic parkas with hoods was not too romantic either.

We did have a golf course in east Montana! It was open only in July and August. The caretaker did not have to worry about keeping the greens cut at the right height as they were coated with sand that had been treated with oil. Usually a golfer would merely add two strokes to his score once he reached the sand green. That worked until someone was able to hit an approach shot within a few inches of the hole and could beat the two-putt rule. Another rule we observed was if one attempted to putt out and took more than two strokes, he was obliged to count each putt.

We had mosquitoes in Glasgow! And were they ever bad. They were so bad that we often wished for an early arrival of winter. An aunt flew in from Kentucky to visit us, and upon her arrival at the city airport, her stockings soon became black with mosquitoes. Without thinking she swatted them and wound up with bloody legs. Another welcome to Glasgow!

Some Crewdogs did like to fish and hunt, so while at Glasgow they were in seventh heaven. I remember one gunner who owned a nice pickup truck. For Christmas he received a gift of a deer rifle with a telescopic sight. He had great fun in aligning the scope and preparing for his first kill. One day he spotted a deer, so he hurriedly braced himself and aimed his rifle across the hood of his pickup. Unfortunately he shot two holes through the hood because he forgot to allow for the scope which was mounted above the barrel.

I lived in four-plex housing on base. It was almost new and quite comfortable. We had a basement and upstairs bedrooms. Our garage was at the end of the building. We were located 5,279 feet from the elementary school. The rule stated that if you lived within a mile of the school, your children were forbidden to ride the school bus. It became the parents' responsibility to get their kids to school as best they could.

So my wife watched the bus pick up neighboring kids just up the street while she struggled to raise the garage door, shovel the snow, start the car and drive over usually slippery roads to the school. Good old SAC had rules that were rigidly enforced regardless of the temperature extremes!

Scheduling Crewdogs for vacations always constituted problems for the staff. If one had kids in school, naturally he wanted a leave during summer vacation. No one wanted to travel the long distances to and from Glasgow in the perils of winter. But there was no way all crews could get choice times for annual leaves. Guess who usually got the short end of the stick? Of course, the newest crews! Such as mine.

One Labor Day weekend my crew got off alert. My wife had checked out a tent and equipment to go camping. Before we could leave Glasgow, she had to judge a flower show at the county fair, so we got a late start for a national park in North Dakota. When we arrived there about 11 PM, the park ranger informed us that the campground was full up. However, he said we could pitch our tent in the overflow area. As that was our only alternative at that late hour, we decided to do just that. We first ate our dinner via headlights on the front bumper of the car at midnight. Then we unfolded the package checked out from recreation supply. Guess what? No tent poles! Not to worry-- as we soon spotted a big tree with low hanging limbs. We were able to use ropes and string and tied the top of the tent to the limbs above. It worked fine. Another surprise was discovered at sunrise. We were camped in the middle of lush poison ivy—and all three of us were allergic to the stuff. By then I was really convinced that only super humans could survive in that forsaken country!

We discovered some very special people while we lived in Montana. Not only did we make many good friends who were in the military, but we also made lifelong friends with some of the civilians in the small town. Since there was a small church of our denomination that worshipped in a barn-like building, we drove the 38-mile round trip from the base on most Sundays when I was not on alert or flying. Those folks never allowed the weather to interfere with their services. I remember one Sunday morning when we were running a little late in getting to church, and I was speeding. I hit a slick spot in the street leading down into town and promptly did a 360° skid. I kept on driving as if nothing had happened, but I headed for the rest room upon arrival.

My wife had worked in a university agriculture and home economics profession before our marriage. So she made friends with the university extension agent in Glasgow. This lady was a charming person who never married and who visited our house regularly. Her sister, also a spinster, was a practical nurse who specialized in using a vibrator and giving rubdowns for sore backs. On several occasions we spent overnight Thanksgivings with them at their log cabin retreat on Fort Peck Lake. After I retired many years later, those two ladies both in their late 80's drove their pickup camper all the way from Montana to Kentucky to visit with us.

While we did at times suffer hardships at Glasgow from isolation and weather, we have many fond memories of the people and the place. Several military reunions for all Glasgow personnel have been held at various locations. Not many other USAF installations have attempted base-wide reunions like those who formerly served at Glasgow, Montana.

Wurtsmith AFB Crew E-37 (L-R) Col Pat H. Earhart, 379th BW Commander; Capt Dick Heitman, A/C; SSgt Ed Osborne, G; Capt Al Singletary, CP; Capt Ron Loy, RN; 1 Lt John "Ray" Houle, N. The picture was tTaken during the presentation of the Distinguished Flying Cross and 8th Air Medal to SSgt Osborne for action as an aerial gunner aboard an AC-47.

Career Crewdog

Dick Heitman

Having been a SAC Crewdog for over 17 years of a 20-year active duty career has taken on a new meaning as I read the B-52 Crewdog tales in the first two Crewdog books. Reading these stories has brought back a flood of memories of flying the BUFF for over 10 years, from 1963 to 1974. Many of the authors' names are familiar, though their faces have become faded in memory over the past 35 years. There are still a few of us around who were Crewdogs in SAC before the B-52 came along. Proud to say, I am part of that group.

My first SAC crew assignment was as a KC-97 navigator at Castle AFB, California in April of 1955. The first B-52s began arriving at Castle AFB in June of 1955, replacing the B-47s in the 93rd Bomb Wing. Little did I know, or even suspect, at that time, that eight years later I would be back at Castle AFB attending the B-52 Combat

Crew Training School (CCTS) as a pilot. The end result was spending the last 10 ½ years of my career as a B-52 Crewdog - never spending a day in a non-crew staff job.

A number of fortunate circumstances during my high school years helped me make the decision to join the Air Force through the Aviation Cadet Program. The minimum requirements to be eligible for the program were to be at least 19 years old, have completed two years of college and be unmarried. I applied for the program shortly after my 19th birthday, passed all tests to qualify for the program and received a draft deferment until I received a class assignment. In January of 1954 I arrived at Lackland AFB, Texas for entry into the Aviation Cadet Preflight Program. Upon satisfactory completion of 12 weeks of basic military training, individuals were assigned to a training base for their chosen specialty - pilot or navigator. Having chosen the navigator route, I went to Harlingen AFB, Texas for the formal navigator training program as a member of Aviation Cadet Class 55-03. With the training successfully completed, I received a commission as a 2nd Lieutenant and the silver wings of an Air Force navigator on 22 March 1955.

Next stop: Castle AFB and assignment to the 93rd Air Refueling Squadron (ARS) as a KC-97 tanker navigator. This was also my introduction to the Strategic Air Command (SAC), in whose clutches I would remain for the rest of my career, except for one year while attending pilot training. Five of us brand new "brown bar navigators" were assigned to the 93rd ARS in the spring of 1955. Training to become a combat-ready crew navigator was started immediately. Air refueling and rendezvous procedures were practiced on almost every flight. Near perfect celestial navigation abilities were also a must, as SAC required its tanker and bomber crews to be able to go any place in the world by using the sun and stars. Being a tanker navigator was a full time job in those days. It was also a time when the navigator could literally tell the pilots "where to go and when to do it", especially when the flight took the pilots out of sight of land or out of range of VORs. Words and acronyms like; Alert, EWO, SAC/Tac Doctrine, SAC IG, ORI, DEFCON, TDY, Stan/Eval, checkride, etc., were becoming parts of our everyday vocabulary. Another requirement before becoming combat-ready was completion of the USAF Advanced Survival Training Course. So it was off to Stead AFB, Nevada for three weeks during the summer of 1955 for hiking through the mountains and living on pemmican bars.

Soon after I arrived at Castle another event happened that put Castle AFB into the AF history books. In June of 1955, the B-52 went operational with SAC when the first aircraft was delivered to the 93rd Bomb Wing at Castle AFB. At that time, the thought of ever flying one of those eight-engined monsters was the furthest thing from my mind. I saw a lot of them up close at the end of the boom during air refueling or when the B-52 pilot would move in close alongside the tanker for photo opportunities. While prohibited from taking pictures of the new B-52B on the flight line, many of the tanker people have quite a collection of airborne photos.

Never very far from my mind was a childhood dream of someday becoming a pilot. I soon realized that the training and flight crew experiences during my first years in the Air Force could be stepping-stones to my goal. About a year after arriving at Castle the opportunity presented itself and I applied for the Pilot Flying Training in Officer Grade Program. Many of my contemporaries in the squadron were doing the same thing. Once accepted for pilot training, we had about a two-year wait for a class assignment. During this time, SAC started crew training for the KC-135, which would soon replace the KC-97. Those of us who had a future pilot training assignment were ineligible for the KC-135 navigator transition training and reassignment to another KC-97 unit was the order of the day. In early 1957, several other navigators and I made the move to the 22nd ARS at March AFB, California to continue crew duty until a pilot training class entry date was received.

Finally, the long awaited orders arrived! Special Orders Number A-8 directed me to report to Spence AB, Georgia, NLT 4 June 1958 for entry into Pilot Training Class 59-H. I completed primary flight training in the T-34 and T-28 in November, and headed to Laredo AFB, Texas for basic training in the T-33, reporting on 1 Dec 1958. All things went well and in June of 1959 I received the coveted silver wings as an Air Force pilot. My previous training and crew experience as a navigator were highly beneficial to my successful completion of the pilot training program.

Assignments for each graduating class were based on the individual's class ranking and which aircraft and bases were available according to the needs of the Air Force. The #1 man in the class got first choice of aircraft and base, and so on down the list. Our class had a wide choice of aircraft and bases. I don't recall where I stood in the class ranking, but my first choice for assignment was to March AFB,

California in B-47s. Why did I pick B-47s instead of some hot fighter? I have been asked that question many times. The main reason was that I wanted to return to the March AFB-Riverside area for personal reasons. During my one year at March prior to pilot training I had become friends with a number of the B-47 copilots and decided that being on a SAC bomber crew was really not a bad deal. Besides, southern California seemed like a good place for a bachelor AF pilot to spend off-duty time.

PCS orders to March AFB and the 22nd Bomb Wing, with TDY enroute for B-47 Combat Crew Training at McConnell AFB and Forbes AFB, Kansas, were forthcoming. Arriving at March AFB in early 1960, I was immediately assigned to a Select crew for a fast upgrade to combat ready status. The B-47 copilot had to be a "jack of all trades", so to speak. A few of his many crew duties included: celestial observations for the navigator, compute performance data, manage the fuel system, operate ECM equipment, operate the tail guns, operate the radios, order flight lunches and even trying to make a landing from the back seat once in awhile. It wasn't long before I had completed crew training, been EWO Certified and ready to pull alert with my new crew of Lt Col Milt Bender, A/C, and Maj Ken Moeller, Nav, and later with Maj Fred Lincoln as Nav. During this time period (early 1960s) of the Cold War era, the B-47 crews at March were spending about one week in every three weeks on pad alert in the "molehole" (underground alert facility at the end of the runway). In addition, some units were pulling 40-day TDY alert tours at overseas bases. While I was at March, our assigned overseas base was Andersen AFB, Guam. During several TDY tours there, we did training flights in and out of Andersen AFB, getting familiar with the dip in the runway and the immediate gain of altitude after passing over the cliff on takeoff. I mention these two items about Guam because they immediately came to mind about five years later when flying a B-52D into Andersen AFB on my first Arc Light deployment.

In the spring of 1963, I was coming up on three years at March AFB and had logged almost 900 hours in the back seat of the Stratojet. It was time to begin upgrade training to pilot and move to the front seat. However, this was not to be, as this was also the time for the 22nd BW to begin transitioning from the B-47 to the B-52. Some crews were sent to other B-47 bases intact. My crew was disbanded because the A/C and Nav were going to retire soon. Looking back, I'm sure the SAC Aircrew Assignments people viewed us experienced B-47 copilots as prime candidates to move into the right seat of a new B-52 crew. My orders soon arrived, indicating reassignment to the 379th Bomb Wing, Wurtsmith AFB, Michigan, with TDY enroute at Castle AFB, to attend the SAC B-52 Combat Crew Training Course with the 4017th Combat Crew Training Squadron. I arrived at Castle AFB in early April for the standard B-52 copilot training program, to include the G/H Difference Course. So there I was, signing in at Castle as a B-52 pilot, exactly eight years after I had first checked into Castle as the brand new tanker navigator. After completing the training curriculum of academics, simulator rides and 10 flights, I departed Castle AFB in July of 1963 for Wurtsmith AFB, Michigan. That was the beginning of almost 10 years of B-52 "Crewdog" duty, taking me to the end of my career and retirement in January of 1974.

Wurtsmith AFB would be my first and only assignment to one of SAC's "Northern tier" bases. Located on the eastern side of the lower peninsula of Michigan on the shores of Lake Huron, the location was sometimes referred to as the "Banana Belt" when compared to SAC's other northern B-52 bases. The base was adjacent to the small town of Oscoda, noted mostly as a summer vacation and tourist spot with good hunting, fishing, and other outdoor activities year around.

Wurtsmith was home to the 524th Bomb Squadron and the newest model of the B-52 then in service, the B-52H. These aircraft were built in 1960 and 1961. When I started flying them, they were practically new; only two-three years old and with only several hundred hours of flying time. For those who don't know, these are the same B-52s flying today (in 2007), with system upgrades of course, and programmed to be in service until 2040. Put immediately into the copilot combat-ready training program, I found it relatively easy to acclimate to my new crew duties. With a crew of six on the B-52, us old B-47 copilots were able to hand over some of our previous crew duties. We now had two people to do the navigation-bombing duties, an EWO to handle all the ECM work and do the celestial shooting and a gunner to operate the gunnery system and order flight lunches. Such a deal! Most of our training on weapons, Tac Doctrine, alert procedures, air refueling, low level navigation, high and low altitude bomb runs and general crew coordination was just a refresher course from B-47 days. There was still plenty to keep a new copilot busy though. New aircraft systems and emergency procedures had to be learned, new EWO targets studied and briefed, in-flight operation of the AGM-28 "Hound Dog" missile and complete familiarity with local operational procedures, just to name a few.

Flying training missions and pulling alert was pretty much the name of the game during my years at Wurtsmith. In 1966 and 1967 the wing was assigned an airborne alert mission known as Hard Head. My Form 5 shows I logged a number of these missions which were big orbits over western Greenland to monitor the Ballistic Missile Early Warning Site (BMEWS), a radar facility north of Thule AFB. Of more significance to me at the time was completing checkout and upgrading to Aircraft Commander in April of 1966. I well remember my first crew and all the help they provided to get this new A/C settled in the left seat and go about the business of flying and pulling alert. The other members of that first crew, E-37, were: Capt Al Singletary, CP; Capt Ron Loy, RN; 1st Lt Ray Houle, N; Capt Jerry Kessell, EW; and SSgt Ed Osborne, G. About that time, some new and different subjects were appearing in our daily intelligence briefings on alert. We were first hearing of Vietnam, SEA, Arc Light operations, contingency procedures, iron bombs, etc. Changes were in the offing for a lot of us at Wurtsmith. Inevitably, crew changes occur for one reason or another. In the later part of 1967 we had a crew change in the RN position and Maj Jim Condon was assigned to the crew. I mention this particular crew change because Jim would end up on my crew again some four years later at a different base, but leave that crew in a very sudden and unfortunate way.

In the spring of 1968 I was coming up on five years at Wurtsmith AFB and very eligible for a PCS assignment under SAC's north/south policy. This resulted in people at the northern B-52H bases going to the southern B-52D bases and vice-versa. And which bases were pulling the load of the Arc Light deployments? In May of 1968 I received PCS orders to the 736th Bomb Squadron, Columbus AFB, Mississippi. I was immediately assigned a crew, had half-a-dozen familiarization/training flights in the "D" model and prepared for a unit deployment to Guam on July 1st. My crew got to fly one of the planes over because I had to finish my checkout in the "D" by accomplishing a heavyweight air refueling on the way over. Coming into Andersen AFB at the end of this 16-hour flight was when I remembered (from my B-47 days) how the dip in the runway could goof up your landing if you didn't allow for the down-slope during landing flare. The other members of the crew were very experienced in Arc Light operations, having completed one or more unit deployments. I thank them for their experience and guidance in getting me through my first Arc Light deployment and completing 74 missions while flying from Andersen AFB, Kadena AB and U-Tapao RTNB. The other members of crew E-21 were: Capt Edward Turner III, CP; Capt Richard Pannier, RN; Capt

Clyde Gray, N; Capt Chester Gonsowski, EW; and MSgt Thurman Lowe, G. The unit returned to Columbus just before Christmas, 1968. At that time word came down that by the middle of 1969, Columbus AFB would become an Air Training Command base and all SAC people would be transferred. Where would we be going?

SAC's criteria for reassignment of crewmembers from the 736th Bomb Squadron were many. I fit in the category of "if a short timer in the B-52D, you'll go to another 'D' base". March AFB was on that list, my request to go there was accepted and so back to the 22nd Bomb Wing for a third time. I returned to March in late July of 1969. The 22nd BW was then known as a "super wing", having two complete B-52 squadrons. These were the 2nd BS and the 486th BS, with B-52C, D, and E type aircraft assigned. Aircraft checkout and EWO certification with a 486th BS crew happened immediately. I had completed the unit Instructor Pilot (IP) training and checkout program prior to leaving Columbus. Shortly after arriving at March, the Wing sent me to the B-52 Central Flight Instructors Course (CFIC) at Castle AFB. This was an excellent training course for moving back to the right seat as an IP. CFIC completion also facilitated my assignment to the Wing B-52 Standardization Division on Crew S-59.

Arc Light deployments were an ongoing event in the life of a B-52 crewmember. March was not under a unit deployment plan in 1970, but rather provided crews on a continual rotation basis. Crew S-59 got the call in June to go Arc Light. I believe this was the second tour for all of us except the navigator who was on his first trip. The crew was: myself, A/C; Capt Rayner Graeber, CP; Capt Paul Bragdon, RN; 1st Lt John Kaschak, N; Capt Vince Osborne, EW; and MSgt D.L. McEntire, G. We successfully and safely completed 49 missions and returned to March AFB in late September. After some leave time it was back to a full schedule of giving standboard check rides, flying crew missions and pulling alert. Toward the end of the year, I managed to schedule a few days off so I could get married. With all factors considered at the time, we decided to have a very small private ceremony and get on with our new life. It didn't take me long to realize that my responsibilities had increased tremendously. Now I would experience how the other married members of the crew coped with family responsibilities while on alert or extended TDY deployments.

The New Year started out with a full schedule of flying and alert. The 22nd Bomb Wing was gradually reduced in size from two B-52 squadrons to one squadron, the 2nd Bomb Squadron. Following organizational and crew changes in the summer of 1971, I was assigned as Chief, Standardization Division, 22nd BW, Crew S-01. The Division was composed of three bomber crews, three tanker crews and two people from Base Flight. Arc Light and Young Tiger deployments were part of our regular schedule and crew S-01 made the list for a 120-day trip to the beautiful island of Guam, starting in March of 1972. We were also told we would be assigned duty as the 8th AF B-52D Tac/Eval crew. After asking some questions, I found out this job would entail flying fewer missions as an integral crew, but involve us in giving written emergency procedures tests to the line crews, flying over-the-shoulder missions with some crews and then briefing the 8th AF Commander on overall results. As a crew, we figured we could handle this job for four months even though we would have preferred to fly as a regular line crew during this trip.

Shortly before our scheduled departure date to Guam, the crew RN received PCS orders for an AFIT assignment. A frantic search for an instructor RN for our crew ensued. Maj Jim Condon, my crew RN at Wurtsmith AFB in 1968, had recently signed into the 22nd Bomb Wing and was awaiting a staff assignment to the Wing Bomb-Nav shop. He had just completed a PCS assignment at U-Tapao in the wing's Arc Light Bomb-Nav office. However, there were no job openings in Bomb-Nav at March and he was going to have to go back on a crew. Knowing his past performance as a top-notch bomb-aimer in B-52s, I offered him the RN job on our crew, providing he could accept the upcoming four-month Arc Light trip. He considered, accepted, and within 60 days he was recertified, standboarded and officially assigned to crew S-01.

The early March deployment date arrived and crew S-01 was on the way to Guam. The crew consisted of: myself (now Lt Col) as A/C; Capt. Ray Hirsig, CP; Maj Jim Condon, RN; Capt Tom Tyson, N; Capt Kent Fry, EW; and MSgt. Don Murphy, G. In February, Operation Bullet Shot kicked off and the lives of many B-52 Crewdogs changed drastically from that point on. For March Crew S-01, our 120-day Arc Light tour as the 8th AF Tac/Eval crew lasted for 20 months (including three 28-day breaks), until 29 October 1973 when all of the 22nd Bomb Wing crews returned to March AFB.

The author at U-Tapao.

Much has been written about Linebacker II already. I would like to add a short personal story about that operation. Our crew flew on the first night in Wave II as Blue 2. Radar problems forced the RN and Nav to troubleshoot the equipment all the way to the target area. We were able to release the iron on target with our cell and safely return to Andersen after a long 14 hours in the air. Our Tac/Eval duties dictated that the crew would perform ground duties with wing staff on future nights (target study, EW and G briefings, command post duty by pilot and copilot). On the afternoon of 27 December we received a call from crew scheduling that our RN, Maj Condon, was to fly as replacement RN on a crew which just returned to Guam from their break, but minus their RN. After verifying all this, Jim got his gear together and went to fly with the crew. Cobalt 1 was lost that night over the target. I was in the command post that night and when the post-strike reports came in I heard the bad news. Jim and his family had been close personal friends of mine since our days at Wurtsmith AFB. And to get him back on the crew when he had returned to March AFB earlier that year, I had told him this Arc Light tour would be short and easy. I hope he has forgiven me for dragging him into that mess! Fortunately he survived the bailout and three months at the Hanoi Hilton. I had the pleasure of going to Clark AB when the B-52 POWs were returned and was able to see Jim the next day in the hospital.

Some crew changes were made for our return to Andersen AFB in early 1973. Our new RN was Maj Jerry Allen, to replace Major Condon. Our 8th AF Tac/Eval duties then turned more to setting up a training program for B-52 crews. There were still some missions being flown, crews and planes were being returned to home bases and things were generally winding down. My concern was about getting back to March AFB by my retirement date of 31 January 1974. The 22nd Bomb Wing crews returned home at the end of October of 1973. That gave me time to get acquainted with my family again, out-process for the last time and complete 20 years and three days of active duty, almost all of that as a SAC "Crewdog".

I wrote this story to pay tribute to all the Crewdogs I ever flew with, to all the ground support people that got our planes into the air, and to a few true commanders that guided me through my career.

> *"If you can learn to fly as a Lieutenant and not forget how to fly by the time you're a Lieutenant Colonel, you will have lived a happy life."* (Author unknown)

THE BEST IN THE WEST
CREW S-59
22ND BOMB WG, MARCH AFB, CA

INTERNATIONAL BON VIVANTS
SOLDIERS OF FORTUNE
LONG RANGE ARTILLERY SPECIALISTS

DICK, RAYNER, PAUL,
JOHN, VINCE, MAC

MEMBERS : VC FOR LUNCH BUNCH
NEWLAND BINGO ASSOCIATION

A crew "Business" Card.

The USAF and Me

Gerald "Jerry" Channell

I was commissioned into the USAF in July of 1957, after graduating from college and completing summer camp in Vermont. My wife pinned my 2nd Lieutenant bars on me. I was happy because I had started what I thought would be a flying career in the USAF. Even as a small child, I had always wanted to fly. My dream was finally coming true! Little did I realize what was in store for my family and me in the nine years I served!

I entered pilot training by way of Lackland AFB in San Antonio, Texas in November of 1957. I flew T-34 and T-28 at Bartow AB, Florida and T-33 at Greenville AFB, Mississippi. I had my first flight in the T-34 on January 27, 1958. It was 17 minutes of sheer delight! I was told to work hard in the academics and flying, because the higher the class standing, the better the assignment. Just before receiving my wings at Greenville AFB, a B-47 landed and sat as a static display for a couple of days. Because I had worked hard and believed what "they" said, I was not anticipating flying one of those things! How wrong I was! 99% of the class went to B-47's, with by-name assignments. This would be one of several times "they" didn't tell the truth. The Officers' Club was not a happy place that evening.

After completing academics at McConnell AFB, Kansas, survival training at Stead AFB, Nevada and flying training at Forbes AFB, Kansas, I finally arrived at Lockbourne AFB, Ohio, in December of 1959. While at LAFB, I applied for several new assignments, but was always turned down: "No suitable replacement available." In late October of 1962, I received a telephone call from my squadron telling me to pack a bag for a week and report to the ready room. That was how I was notified of the beginning of the Cuban Missile Crisis. I spent eight days on alert at Philadelphia International Airport, living in a motel and wearing my flight suit and my 45. It was a sobering experience.

Knowing that the B-47 was going to the "boneyard" soon, I finally volunteered for B-52 training. I would miss my marvelous Aircraft Commander in B-47's. He was also an Instructor Pilot and let me fly in the front seat. I had completed most of the upgrade requirements, including refueling, when I left for B-52 school. I was tired of the alert cycle at LAFB and the three-week TDY "Reflex" cycle to England, which the B-52 didn't do. So late in April of 1963, off to Castle AFB, California I went with my wife and three children. "They" once again failed to inform me of the three-year commitment I had incurred when I accepted the assignment. "They" didn't tell many, because in talking with some other pilots, I found they were surprised also. This three-year commitment would haunt me in 1964.

We spent two weeks sightseeing across the country, finally arriving in Merced, California in May. It was difficult to find a place to rent for the three-month TDY and with a wife and three kids it was even more difficult. Shortly after moving into the first place we rented, the septic system failed and we had a smelly pond in the backyard. We left that place and rented another one that had an irrigation ditch at the end of the backyard. No fences either! Once, we even had to live in a trailer! Such was life if your dependents went with you on the TDYs. After completing academics, I had my first flight in the B-52F on June 10, 1963. Training was easy at Castle AFB and on July 1, we departed and drove to Amarillo AFB, Texas. Since I arrived at AAFB as a spare copilot, I did most of the training for combat crew duties alone. I might add that we also bought a nice house on the west side of Amarillo; the base was on the east side. I passed all the checks and was assigned to a crew and pulled my first alert in a B-52D on October 7, 1963. Alert duty was not new to me because of my three years in the B-47.

I was on alert on November 22, 1963 when President Kennedy was killed. If I remember correctly, the "Klaxon" went off, we all ran to our aircraft and started engines. We received a message to "shut down engines and stand by." None of us had ever received a message like that from the Command Post. I can't remember how long we sat in the aircraft, but it was long enough for all of us to do some wild thinking.

After a while we were released from our planes and once we were back in the alert barracks, we learned what had happened. All of us spent the day watching TV!

I flew and pulled alert with that crew for about a year and a half and was then assigned to an Instructor Pilot for the purpose of upgrading to Aircraft Commander. My IP was a great guy and I had lots of good times with him. He made upgrading enjoyable. He even christened me with the name of "Gorilla Grip" Channel! He called me that because in the early stages of air refueling, he tried to help me smooth out on the control movement before power steering came in. I was holding on to the yoke so firmly, he couldn't move it. "What a gorilla grip you have," he said, and I was stuck with that name. In fact my crew gave me a memento at my farewell party of an eight-inch gorilla doll holding on to a control column with a B-52 control wheel medallion in the center. The nameplate on it says "Gorilla Grip Channel!" I still have it; it sits on my desk reminding me of my crew days.

I had only one refueling with the KC-97 in the B-52, and it was a mess. I was number one in a two-ship cell and we were refueling with three KC-97s. It took about 30 contacts to get 15,000 pounds of fuel from the number two tanker, but only one contact for 30,000 pounds from the number one. I never could figure out what happened! I had refueled with a KC-97 in a B-47 several times and never had that much of a problem.

I was becoming more and more disillusioned with the Air Force because of the alert duties, the hopelessness of getting out of B-52s, the 20-year mark and possibility of forced retirement happening when my kids would be in college and my expenses high. The Vietnam situation was heating up and I traveled to Dallas in November of 1964 to have an interview with American Airlines. My interview, physical, and simulator ride were successful and they offered me a class date. Upon submitting my resignation, I found out about the three-year commitment for the B-52 training. "They" were at it again. The earliest I could get out was fall of 1966. To top it all off, they announced that Amarillo AFB was closing and the bottom fell out of the housing market. We "gave" away our house in December of 1966 for $50.00 to a nice civilian couple. They got a great deal - we got the usual AF deal.

In April of 1965, I finally completed my upgrade training and got my crew, E-07. So off to the alert barracks, "the Mole Hole", I went. Of the 1,230 days I was assigned to the B-52D, I spent 376 days on alert and another 99 days flying. I was away from home two out of every five days. That didn't include the days we mission planned or days of "special" meetings at the squadron. I also got called in to explain why I didn't have a savings bond deducted from my paycheck, why my navigator didn't mow his lawn, why my wife didn't belong to the Officers' Wives' Club, and why I was resigning. My mind was pretty well made up.

On the whole, my three years in the B-52 were quite uneventful although there were a couple of incidents that I remember quite vividly. Our squadron flew the 24-hour "Chrome Dome" (airborne alert) missions several times, in addition to the regular ground alert. This mission took us from Amarillo AFB, over Maine, along the East Coast of Canada, straight north to the North Pole, over Alaska, out to the Aleutians, back to Seattle and straight to Amarillo. We had two refuelings, one for 100,000 pounds and one for 120,000 pounds. I flew eleven of those flights. Most of the time we had just the basic crew for a 23-hour flight and no extra Pilot or navigator. My EW became very proficient at flying the autopilot! He was also a great waiter on those long missions. After eating the in-flight kitchen meals a couple of times, my crew decided to try something new. We all put money into a pot to buy frozen meals at the local supermarket. My EW was put in charge of this program. For the remaining Chrome Domes, we had hot dishes served by a gourmet waiter with a white cloth on his arm!

One of the incidents I remember very well was the crash of a KC-135 from the refueling squadron at Amarillo. It had crashed while shooting touch-and-goes at the base. There were no survivors. I was flying that night and I believe that it was the tanker that I had refueled with. As we approached the Amarillo area, we were instructed to proceed to Dyess AFB at Abilene, Texas and remain there overnight. We flew back to Amarillo the next day and saw the remains of the aircraft. It was a sad day for all. It was more difficult for me because my mother and father were visiting with us. They were quite shaken up!

My crew got a passing grade on a Combat Evaluation Group visit. We did our preflight of the aircraft, started the engines, taxied out, and turned the corner for a rolling takeoff; pushed the power up and both of the front windshields shattered. Obviously we aborted the takeoff, taxied back, shut the engines down and the flight was cancelled. Because we had applied power for takeoff, it was considered a flight, even if we didn't leave the ground. SAC was strange!

The incident I remember the most was during one of the Chrome Dome missions. We had just passed over Thule, Greenland, when I smelled smoke. My copilot smelled it also. There is nothing like smoke to get the hairs on your neck standing up - nothing worse than a fire in an airplane. We quickly checked all our instruments to see what might be the problem, but everything checked normal. We all could still smell the smoke. My EW left his station to check the area between his station and the pilot's seats and relayed that the smoke was stronger in that area. After a short time, he found the problem. The Aldis Lamp had fallen from its clips onto the jackets in that area. Something heavy had been placed over it and had enough weight to turn the trigger switch on. My flight jacket had a nice three inches round hole burned into it. Needless to say, we were very careful where we stowed equipment from then on.

I re-submitted my resignation in September of 1966, and shortly after that I found out that my squadron was going to Guam/Thailand in early 1967. It was interesting to fly the "Iron Bomb" practice missions. I remember dropping bomb shapes on Matagorda Bombing Range. It is no longer a bombing range and I live fairly close to it today. I also got to fly formation, not close, but close enough for the "Aluminum Overcast."

I've read the stories in the other two "Crewdogs" books and I must say that the term "Crewdog" was not used by me or by anyone else that I knew in all my years in SAC. I guess it must have become prevalent in the SAC vocabulary after I tendered my resignation of my Regular commission. I have a few friends that left the B-52 around the same time I did and they can't remember the term being used either. It might be interesting to find out when and where it originated.

The other term that is used a lot in the stories in the books is "BUFF". I remember the term "Aluminum Overcast" being used. I'm not sure that "BUFF" was around or used very much when I left SAC. I can't remember using it in Amarillo. However, one of my friends who flew B-52's at Roswell AFB, New Mexico, stated that he had heard the term, but it wasn't used very much. Most of the writers use the word quite frequently in their stories. In the second *"We Were Crewdogs"* book, I read the story that attributed the term to CINCSAC at Kelly AFB, Texas in 1966. This makes sense because during 1966 we were receiving the B-52's that were painted black and camouflage colors.

While the AF taught me how to fly, I believe they got their money's worth out of me - especially in the last couple of months of my service. I was always the A/C they used to replace the A/C on alert who was sick or had some other problems. All the alerts, ORIs, night flights, "Chrome Domes", and TDYs (mostly in the B-47) that I performed certainly gave the Air Force a very good return. Such was the life of a crewmember during the "Cold War" era - from one alert cycle to the next and maybe flying in between.

My reasons for leaving were pretty standard: closer family life, could make more money as a civilian, my interest in a career in the AF ceased to exist, and I had a job with American Airlines.

Twelve days after I resigned, I went to work for American Airlines and spent 29 years flying for them. My one claim to fame is that in 38 years of flying, both military and civilian, I never put a scratch or dent on an airplane.

Exigencies of the Service

George R. Thatcher

Fresh from Undergraduate Pilot Training and B-52 CCTS, I arrived at Turner AFB, Georgia in late 1960. From a pretty fair variety of potential flying assignments, I had chosen the BUFF for a couple of reasons. The first one was my gullibility in believing the glowing propaganda of a senior SAC officer sent to Vance AFB to "counsel" us newbie UPT graduates. This guy told us that we'd soon be spot-promoted as combat Crewdogs, and that we'd only be filling crew positions for a three to five year tour, after which we'd be off to greater glory as staff officers.

This leads to the second reason I chose the BUFF. I was one of the last of the "mustang" officers, having graduated from Officer Candidate School after almost six years in the ranks and pulling a subsequent short tour as a radar intercept director while awaiting my assignment to UPT. To a guy who came on board with no college, the prospects for augmentation into the Regular Air Force and a respectable career would seem to have a lot to do with getting that spot promotion and being able to finish college on a "bootstrap" assignment after a few short years of crew duty.

I soon found out about promises made by SAC weenies and other snake oil peddlers. Spot promotions were available IF you were a second lieutenant or IF you happened to be assigned to a "Select" crew - neither of which applied to this FNG in the squadron. But as an eager young low-time pilot, I decided to get my head into this business and become the best I could be at what I did. As time went on I slowly built up toward the required 2,000 hours of flight time before upgrading to aircraft commander. In those days this was a tedious process requiring the better part of five years in the right seat. I learned that my formal education was to be put on "eternal hold," owing to the exigencies of the service that made me the most valuable Crewdog in the squadron, if not in all of SAC.

When the time came for me to begin the upgrade process, I was able to cadge some extra left-seat time by volunteering to fly extra missions, some of them being the 24-hour "Chrome Dome" airborne alert variety. On one of these I was assigned to fly with Capt (later Lt Col) Walter Black, who was a newly-certified aircraft commander. The "powers" figured he could use a high-time copilot along for balance, until he got a few missions under his belt.

Everything was normal during our first several hours of flight, and then it came time for our first air refueling over the Mediterranean. It was late at night and our refueling track was in an area of high cirrus clouds and continuous light, choppy turbulence. As I recall, the first refueling load was scheduled for 120,000 pounds. Walt and I agreed that he would hook up with the tanker, and after getting stabilized and beginning to take on fuel, he would let me operate the throttles for fore-and-aft control. It was a procedure that had served me well through the past several years, but this night was different.

As Walt "parked" our bird in position and began to onload JP-4, I took the throttles as agreed. After a few minutes of coping with the choppy air and Walt's increasing nervousness, we disconnected. The second attempt didn't go much better, and I found that I was having to use an awful lot of throttle movement to keep us in the refueling envelope. In about 10 minutes time we took on about 30,000 pounds of our scheduled fuel and poor Walt was sweating like the proverbial hooker in church. But he kept on trying, and kept on disconnecting. Finally, after one particularly wild pass at the tanker, he said to me, "Take it, George, I'm going straight down."

I knew instantly that Walt was experiencing severe vertigo. Given the pressure to perform on his first major trial as an A/C, the choppy air and the blackness of the night, it wasn't hard to understand why. Additional pressure came from the fact that if he didn't get the gas, we'd have to divert to an alternate base in North Africa. Walt's future career, if not his whole life, was passing rapidly before his eyes.

As we backed off behind the tanker to re-group, Walt asked me if I thought I could finish the refueling. Now I'm not a mind reader, but when I saw the desperation in Walt's eyes as he asked that question, I knew I couldn't let him down. We briefed the tanker that there would be a "change of crew positions" and went back in, this time with me at the controls and Walt running the throttles after we were safely tucked in behind. We got the gas, and repeated the same procedure at our second refueling. Walt and I kept the experience as nobody's business but our own for many years.

About 30 years later, I attended my first Arc Light/Young Tiger reunion in Dallas. By then I had begun writing poetry about my flying experiences, and one of them was about air refueling. I was hoping that Walt Black would be at this reunion so I could present him with a copy of it. As it happened, he was there, and he spotted me coming in the door of the reunion hotel. With a big smile on his face he called out to me and said, "There's my savior."

I don't know anything he could've said that would've made me prouder to have been a BUFF Crewdog. Walt, one of the finest men I ever knew; passed away recently.

Midnight Air Refueling

George Thatcher © 1998

It's hard to imagine a job as grueling
As a max-weight, midnight air refueling.
Close formation is just a lark,
Compared to gassing in the dark.

Two rows of colored, blinking lights
In dazzling contrast to the night,
Are all you have to guide your flight.
Move up, back off, swing left, now right.

Young pilots' insides turn to jelly,
When scraping against a tanker's belly,
The task becomes foreboding, frightening,
You're trying to bottle streaks of lightning,
As first you try to chase the blinks,
Out of sync with the jukes and jinks.

The lights entice, they're teasing you,
Too low, pull up, don't overrun!
Dear Lord protect us, prays the crew,
Disorientation in everyone
Aboard the bird; your inner self
Says, "Help, I'm falling off a shelf!"

The sweat runs down the lines and creases
Of your face, as the weight increases.
Fatigue becomes an aching moan
In every sinew, muscle, and bone!
Radio discipline, can't even ask
The boomer if we've got the gas.

Twenty minutes on the boom,
Twenty-five, you're soaked in sweat,
Tanks can't have that much more room,
God, why aren't we finished yet?

Finally, the disconnect,
Copilot quickly reads the gages.
Lights fade off into the night,
Tonight, by God, we've earned our wages!

Catch 22 ReDeux

George Schryer

In late August of 1969 as I processed though Personnel and Finance offices at Seymour Johnson AFB, North Carolina, in preparation for my first trip to the mysterious Orient and the land of Charlie Towers and Shark Pens I tried very hard to make all the right decisions. By that time in my military career I had made a temporary move more than once so I knew how things could change when least expected. I was leaving my wife of only eight months and her daughter and I wanted to make sure that they would be all right while I was gone. We were living on base, surrounded by fellow Crewdogs, and her family was living up the road in Greenville about 45 minutes away so I felt she would be all right. But just to be sure I told all the processing specialists "Don't change a thing. Keep all my records and pay just the way it is now and I'll straighten it all out when I get back." Little did I know! I think the phrase is "FUBAR."

The time finally came and we said our good-byes and off we went. After a short stay at Castle to transition back to the D-Model it was on to the lush island of Guam. Guam is good, or was until I received a letter from my wife a couple of weeks later. Basically she wanted to know what happened to my paycheck. It seems as how, in their infinite wisdom, the local finance people decided that when I specifically said don't change a thing I really meant change everything. Instead of giving me my regular pay, flight pay, combat pay and tax exemptions they were sure I must have meant take all those things out of my pay. That allowed my wife $100.00 a month to pay all our bills and send me money to live on.

Needless to say I beat a hasty retreat to the Finance Office to correct this situation only to find out that "Sorry Guns, we don't have your pay records here. They remain at your home station while on Bullet Shot." Okay, call Seymour Johnson and have them straighten this mess out. "Sorry Guns, they can't do anything with out you being there." Okay, how about a partial payment? "Sorry Guns, we can't do that here. It has to be done at your home station." Okay so what do I do? "Gee Sarge, I don't know what to tell you." Finance people Number 10 GI!

Fortunately by promising the car dealer and Sears and a few others our next child and borrowing from friends and family she hung in for the next six months. I was one fortunate gunner as my crew helped me on my end and I was only required to treat them like royalty and wash their feet after every mission. (Not really, but I sure was indebted to them and feel to this day I still owe them.)

We worked into the flow and flew our missions, bought our Guam Bomb, got drunk at Mr. K's and did all the other requirements imposed on SAC's trained killers. Then it was on to Okinawa for the second chapter of our tour of the Orient. Guam was good but Okinawa was muuuuch better. Even on a severely restricted income and eating a 50 cent bowl of fried rice for supper each day I still managed to have a good time. Flying was pretty routine, except on the flight that got me my Bonus Deal, and we fell into the rhythm of flying and drinking.

Shortly after Christmas of 1969 I realized that I was due for reenlistment on the upcoming 4th of January so I thought it might be a good idea to go to Personnel and check on my paperwork. Number 10 GI work in Personnel also! "Sorry Guns, there is no paperwork here for you to reenlist Sarge. That should have come from your Personnel people back at Seymour Johnson, we don't generate that here. The only thing we can do is contact them and see what the deal is." And that will take how long? "Oh we should know within a week, remember this is the holidays and things take longer." Now let me see if I got this straight; after we take off on a mission, but before we can drop our bombs on some monkeys in the jungles of South Vietnam we have to call the White House for permission, but it is going to take you a week to call Seymour Johnson and see if I can reenlist? "Sorry Guns, it really is not quite that simple." Well of course not, this is the Air Force.

Oh by the way, what happens if the 4th of January gets here and there is no paperwork? "Well then Sarge you can't reenlist. We will provide you with a free airline ticket to your last place of enlistment. You will be a civilian then." Talk about your cloud with a silver lining. My last reenlistment was at Lincoln AFB in Lincoln, Nebraska which had closed in 1966 and my family was in Goldsboro, North Carolina, but both places were back in the land of the Big BX. "Oh Guns, just so you know, you can reenlist when you get back to Seymour Johnson."

Decisions, decisions, decisions. Well I figured I had better inform the A/C that he may not have a gunner next week. Needless to say he was not too enthused with my news, probably because I owed him and the rest of the crew a decent hunk of change and keeping me with them insured that they would get paid, or maybe it was 'cause he was just a great guy.

He said "Guns wipe the smile off your face and I'll see what we can do." A couple of hours later he came back and said for me to go to Personnel in the morning and they would have the reenlistment paperwork ready for me to sign. I asked him how high he had to go and he just said there are people with eagles on their collars who know whose button to push and when. He also said with a smile "Contrary to what you have been led to believe Gunners don't have all the power all the time." At least he understands the all the time stuff.

So when 4 January rolls around guess who is on the morning schedule? You bet money don't you? Yuup, we were number three in the second cell or should I say we started there? We ended up bag dragging to tail-end-Charlie but managed to make the parade. After systems checks and Air Refueling the A/C asked me if I was ready. I asked him if we could do it on the Bomb Run but he said "Don't Press it!" So at 35,000 feet over the blue Pacific he (in the nose of the plane) read the oath and I (in the tail, 150 feet later) repeated it while I looked at a small American flag I stuck in my window. A round of congratulations came over the interphone when I finished. I think they were relieved that I would be around to pay them back.

Shortly thereafter we moved on to U-Tapao and things began to look up - mainly because I didn't have any more dealings with Personnel or Finance. Then it was back to Guam and back to the World. A couple of days later I made my presence known to the personnel in Base Finance and it was apparent that an Eagle had flown through the office just before I got there because it seemed like there was nothing they wouldn't do to make my stay in their world as short as possible. In no time at all I was informed that they would have my check cut for me momentarily. But I just couldn't let it go. "I do not want a check," I told them (the amount made me catch my breath) because with my reenlistment bonus it was pretty hefty. I informed them that I wanted the entire amount in 20's.They were quick to grant me my request and even offered to call and have an MP escort me to the bank if that was what I wanted. "No thanks," I said.

When I returned to the house my wife said sarcastically "How long before we will get the check?" I told her they wouldn't give me a check and before she could respond I threw a fistful of 20-dollar bills in the air and showered her with money. After what she had been through she deserved it. The next day we settled up with all our creditors, crew, and friends.

Crewdogs really are family.

The author and his D-Model crew in Guam.

A Warrior's Story – Vietnam to Iraq

Buddy Sims

I received my orders with all the excitement expected of a newly commissioned 2nd Lieutenant fresh out of Air Force ROTC from Washburn University, Topeka, Kansas in June of 1967. I had no way of knowing it then, but my first assignments in Vietnam as a military officer as an 0-2A Forward Air Controller (FAC) in November of 1968 and later as a B-52D bomber Pilot in May of 1972 would reverberate down the corridors of time 30 years later when I directed the time sensitive targeting over Iraq and Afghanistan from December of 2002 to May of 2003.

When I returned from my Vietnam FAC tour in November of 1969, I was sent to Castle AFB, California to qualify as copilot on the 450,000 pound B-52F heavy bomber with assignment to Wright Patterson AFB, Ohio. There I qualified on the B-52H and was assigned to the 17th Bombardment Wing of the Strategic Air Command (SAC). My tour at Wright-Patterson came during the height of the Cold War when the B-52s were on 24-hour alert. The big bombers stood at the alert facility fully loaded, engines cocked with starter cartridges, and ready to fly with just minutes notice. As the third leg of the nuclear triad, we knew that if we did get the execution message, we would see much of America in nuclear ashes as we crossed over the continent toward targets in the old Soviet Union, carrying a formidable load of four thermo-nuclear gravity bombs and nuclear-armed AGM-69 SRAM-A short-range attack missiles.

My B-52H crew from the 17th Bombardment Wing was deployed to duty over Southeast Asia after qualifying in the B-52D at Castle AFB. We arrived at Guam on a direct B-52D flight from the States and landed in the middle of full scale combat flying operations in May of 1972. Charlie Tower didn't even have time to help us find a parking location. My first bouncing landing going downhill with an empty aircraft was something my crew will never forget! I was a young captain but had the advantage of two crew majors - Radar Navigator and Electronic Warfare Officer with previous Arc Light tours. We were assigned to the 307th Strategic Wing for six months, which flew out of Guam Island and from U-Tapao AB in Thailand. We made most of our bomb runs over North Vietnam and periodically based at U-Tapao AB where we would run a week of short range bombing missions before rotating back to Guam to prepare to fly more long range missions. The bombers logged 6,500 miles and 12 hours of flying time in its Guam to North Vietnam round trip. The B-52D always flew in three ship cells and usually was loaded with either 108 bombs from U-Tapao or 66 bombs from Guam. The B-52D was used mostly over North Vietnam due to its better electronic warfare capabilities as compared to the B-52G which was loaded with 27 bombs and less capable ECM gear.

My favorite "Crewdog" story in the B-52D happened when returning to Guam in 1972 during a 10,000 ft bomb bay check for hung weapons. The Radar Navigator was crawling his way past the wheel well to the bomb bay when Guam Approach asked us to expedite our descent – I just happened to be at the relief can and heard the screaming as my copilot, without thinking, dropped the landing gear. The scared-to-death and quite upset Radar Navigator to this day has never forgiven the copilot.

I flew 100 missions over North and South Vietnam in the B-52D and experienced my share of 10-minute bomb runs with the EWO sternly advising me, fire warning lights from hot bleed air leaks, hung bombs, and those challenging air refuelings in marginal weather conditions. Many thanks go out to those tanker crews for all those years of passing gas to us, finding us over the ocean, and especially the dedication and professionalism of tanker boomers.

My crew returned to Wright-Patterson AFB in October of 1972 where I served as a B-52H Instructor Pilot/Evaluator. Next stop was Barksdale AFB, Louisiana, where I completed a three-year tour as a 1st Combat Evaluator Group check pilot. After a tour in England as an Emergency Actions Officer, I was assigned to Grand Forks AFB, North Dakota, as a B-52G/H Instructor/Evaluator Pilot and the Director of Training. There I helped build the new B-52H contingency (conventional strike) force which was started in 1979 and remained essential to the nation's defense. These missions were not without peril and sacrifice, especially getting the pilots to fly night mountainous low level terrain avoidance. This new conventional projection of national will and resolve was called the Strategic Projection Force (SPF) with combined aircraft and crews from Grand Forks and Minot AFBs. During the takeover of the American Embassy in Iran our aircraft and crews deployed to Guam as a show of force and were prepared to strike Iran if given the order. I also flew on a 31-hour non-stop Bright Star mission to Cairo, Egypt from Grand Forks to make a live bomb drop to impress the Soviet Union with our long range conventional strike capability – CINCSAC even gave me an Air Medal for bombs on target and on time!

Another perilous incident that I observed occurred on September 15, 1980, when a fire broke out on a B-52H during an Emergency War Order exercise at Grand Forks AFB. The aircraft was fully loaded with nuclear gravity bombs and AGM-69 missiles. The fire raged out of control for over three hours and resisted all efforts to put out the blaze. The situation became so desperate that base officials were preparing plans to evacuate everyone within a 20-mile radius of the base, including the community of Grand Forks.

There was no danger of a nuclear explosion, but if the conventional explosives in the bombs and missiles were ignited by the fire, there was a great risk that the radioactive plutonium in the nuclear bombs would be dispersed into the atmosphere and cause heavy casualties as well as poison the ground for years.

If it was not for the personal bravery of the military fire chief who went into the smoke-filled cockpit at great personal risk and successfully re-set the fuel cutoff T-handle after two unsuccessful attempts, the worst would likely have happened. What we learned: If the copilot turns off the battery switch before pulling the fuel cutoff T-Handles, the fire will not go out.

After a four year tour in Panama as the Deputy Commander at the Inter-American Air Forces Academy and then two years in Security Assistance in the Pentagon, I retired in January of 1989. I flew commercially as a Pilot for United Express and Pan American World Airways followed by a new career in managing the installation of avionics equipment for the EA-6B Prowler, a carrier based electronic warfare aircraft for the Naval Air Systems Command (NAVAIR).

In 2002, there was a shortage of USAF Crewdogs and under special legislation in the Fiscal Year 2001 National Defense Authorization Act, Congress authorized the services to offer recall opportunities to as many as 500 retired military officers in the most critically needed professions. The Air Force was given slots for 110 pilots, 75 navigators, and 32 air battle managers—217 in all.

I knew of the Retired Aviator Recall Program (RARP), but I didn't think much about it until January of 2002, when I received an email from a longtime B-52D Air Force friend, Lt General Thomas Keck, Commander of the 8th Air Force at Barksdale AFB, Louisiana. His message was direct and to the point: the Combat Operations and Plans Squadron at 8th Air Force needed combat tested staff officers for the training of personnel in the Combined Air Operations Center - Training (CAOC-T) in order to free up A-10, F-16, and B-52 pilots and navigators for combat duty.

After 9/11, I was ready to volunteer and had to pass the same physical examinations as an entering 2nd Lieutenant pilot. After four months and three separate flight exams, at age 58, I reported for duty at Barksdale AFB in May of 2002. I was assigned to the 608th Combat Operations Squadron, which has the task of battle management training for the CAOCs, which are positioned in strategic locations all over the world. We performed the operation's floor execution of the Air Tasking Order (ATO) in war fighting.

In August of 2002, I deployed to a CAOC exercise known as Ulchi Focus Lens being conducted at Osan AFB, Korea and qualified as a staff B-1, B-2, and B-52 bomber duty officer. Following my return to Barksdale, the 8th Air Force received orders in November to deploy as Air Expeditionary Force (AEF) 5/6, to Prince Sultan Air Base, Saudi Arabia, its first deployment into a combat zone since World War II.

After a three-week command and control exercise enroute at Shaw AFB, South Carolina, we arrived in mid-December of 2002 and relieved the current CAOC staff. Their mission was to support air operations over Afghanistan and the southern Iraq "no-fly" zone—Operation Southern Watch. Even though the air war over Afghanistan was winding down and Operation Iraqi Freedom was still months away, the southern Iraqi no-fly zone was an intense combat area and American and British pilots were being fired upon day and night. I served as Chief of Time Sensitive Targeting (TST) with the responsibility of coordinating the Iraqi air war in Operation Southern Watch and Afghanistan. I directed combat response options to counter the Iraqi threats to our coalition forces working side by side in the TST Cell with a B-52 Radar Navigator, another recalled B-1 Pilot, and an RAF Wing Commander.

While in Korea, I had trained with a new software program that General Dynamics developed called the Automated Deep Operations Coordination System (ADOCS) and soon after arrival I was tasked to implement ADOCS into the Saudi CAOC computers. ADOCS provided a powerful suite of electronic tools and interfaces for horizontal and vertical integration across the joint battle space and offered an unprecedented capability to manage and execute war planning. I helped the General Dynamic contractor install, trained the CAOC staff, and put the system into operational status. Using ADOCS, I managed the battle space from a computer screen in a ground based operations center, using technology that I could not have imagined 30 years prior in Vietnam.

RAF Wing Commander Bryan Trace and I directed the first combat strike against an enemy target on January 21st of 2003 using the ADOCS system—a strike against an Iraqi missile control van that had been directing ground-to-air missiles against American pilots patrolling the no-fly zone. I directed some 50 time-sensitive targeting missions during the December of 2002 to February of 2003 timeframe, employing such diverse weapons systems as the F-14, F-16 and F/A-18 fighter-bombers, the Predator drone armed with Hellfire missiles, and the S-3B Viking surveillance and precision targeting aircraft.

On the flight line was also the GBU-43/B Massive Ordnance Air Blast Bomb (MOAB), the largest ever satellite-guided, air-delivered weapon in history. It was over 21,000 lbs total weight GPS-guided munition with fins and inertial gyro for pitch and roll control. The

"Mother of all Bombs" was targeted into Iraq but never dropped – it was really part of the "shock and awe" strategy integral to the invasion of Iraq as described by Secretary of Defense Rumsfield. There were no Laws of Armed Conflict principles or treaties that prohibited the use of the MOAB weapon in Iraq.

In early February of 2003, Katie Couric and the NBC Today Show broadcast live from the Prince Sultan AB flight line. As part of AEF 5/6, we were deployed for 90 days until Katie asked our one star base commander how long the troops would be staying in Saudi. He broke the news that AEF 5/6 was just combined with AEF 7/8 for another 90 days. The entire flight line of 1,000 troops all sighed in unison! So much for notification by personnel.

General Michael Moseley, Combined Forces Air Component Commander, separated the Afghanistan and Iraq control centers in late February of 2003, sending half of us and the Afghan mission to Al Udeid AB in Qatar. A new 60-million dollar CAOC was activated, manned, and continues to this day to direct the fight against terrorism. Following the end of major Iraqi air operations in April of 2003, Prince Sultan AB was closed to the American military and both missions were reunited at Al Udeid AB. I continued to direct time-sensitive targeting and close air support missions over Iraq and Afghanistan until my return to Barksdale AFB in May of 2003.

But before I left Barksdale AFB, I had the opportunity to take a trip down memory lane and once again get behind the controls of the same B-52H bomber (61-017) that I flew starting in 1970 until my retirement in 1989 - it was the last time I flew as an Air Force B-52 Command Pilot in June of 2003. My flight record reflects over 33 years in the B-52H as a pilot. Lt Col (Ret.) Bill Jankowski, USAF, the 0-2A FAC shot down in North Vietnam in 1972 whose story was told in the book and movie *"Bat 21"* was my instructor pilot. To his further credit, he also flew B-52H bombing missions over Baghdad out of Diego Garcia in 2003. This was surely the highlight of my B-52 flying career after being away from the aircraft for many years on the retired list. I ended up with 4,500 hours in the B-52F/D/G/H bombers and served as a CP, P, AC, IP, and EP.

Lt Col Jankowski and the author with 61-017 in June 2003.

I PCS'd from Barksdale to the Pentagon in June of 2003 for my second year of active recall duty and served directly under Major General Steve Goldfein, Directorate of Operational Capabilities Requirements, Deputy Chief of Staff/Air and Space Operations, Pentagon, Washington, DC. One of the Air Staff Divisions I helped manage was tasked to keep the B-52H bomber in service thru 2045 – this means we'll be able to watch it for many years in the future – putting bombs on targets. I retired in June of 2004 and was the last soldier still on active duty who had served in the Cuban Missile Crisis as an enlisted marine and at 60 years old felt it was time to hang up my flight suit.

Back into retirement I had one more goal to accomplish, I helped plan, design, and construct the "Freedom Park Memorial" in Edwards, Colorado, and enlisted the help of the county commissioners, business professionals, representatives from the local police and fire departments, mountain rescue teams, and military veterans who served in WWII, Korea, Cuba, Vietnam, and Operation Iraqi and Enduring Freedom. This Freedom Park Memorial was dedicated in July of 2006. It includes an American Flag Pole Plaza, a 600 lb piece of the actual Pentagon Limestone from the 9/11 attack, inscribed names of Eagle County fallen soldiers, and over 200 memorial pavers. No higher honor in my life was the installation and dedication of the 27 memorial remembrance granite pavers to those B-52 "Crewdogs" who died in Linebacker II in December of 1972. In Edwards, Colorado, I remember and memorialize these brave airmen from the Eleven Day bombing campaign that brought the Vietnam War to an honorable conclusion and returned our POWs. Col (Ret) Tom Kirk, USAF, a member of my VFW Post 10721 and POW at the Hanoi Hilton in 1972, remembers to this day the sacrifices made by the B-52 bomber crews in ending the Vietnam War. Tom said "When we heard the long trails of bombs exploding we knew the B-52 had come to our rescue and our freedom was coming".

We must always remember that we won all the battles in Vietnam, but the U.S. Congress and war protestors lost the War, not us! In Vietnam over 70 percent of our military were volunteers while in Iraq and Afghanistan today 100% are volunteers. The country must always "support the troops" and never allow the military to be disparaged again like during the Vietnam War. From my recall experience and working on a daily basis in the trenches at a "Joint Fight" command center, I can confirm that America is so fortunate and lucky to have such outstanding dedicated men and women in the Armed Forces.

My "A Warrior's Story - Vietnam to Iraq" input is offered to point out that wars continue to be fought by volunteer airmen who desire to serve this great county and especially to highlight the best aircraft ever built by Boeing – the B-52 bomber!

Chapter Two

C**old War** [kohld] [wawr] – *noun* - A state of political tension and military rivalry between nations that stops short of full-scale war, especially that which existed between the United States and Soviet Union following World War II.

Mors Ab Alto

Tommy Towery

"Mors Ab Alto" was the motto of the Strategic Air Command's 7th Bomb Wing, the unit in which I served my stint as a B-52 EW. I don't know why, but I was mesmerized by the meaning of that motto, first heard during an intelligence briefing when I was doing my initial certification for the wing's Emergency War Order. Its meaning was clear and concise and the gist of it has stayed hauntingly in my mind ever since that briefing room revelation. It is the essence of the truth about the reason for the existence of the B-52. It is three simple Latin words that describe what could be the mission statement of all the bomber wings and all the Crewdogs that flew the BUFF. For that matter, it equally has applied to any bomber crew of any bomber aircraft of any war. The Latin words Mors Ab Alto translates in English to "Death From Above."

As I withdrew my hand from the metal and as the last contact between my skin and the behemoth cylinder diminished, I knew that I had just joined a closed brotherhood of rare human beings. I had done something that no relative of mine before me had ever done and perhaps no relative after me would ever do. I had actually touched a nuclear weapon or what in my "duck-and-cover" childhood days, I would have called an "atomic bomb." I had touched one of the most horrible death creating devices ever devised by man. I do not believe that I am the only one that felt the need to perform that action as part of the initial training needed to be completed before during duty as a BUFF Crewdog.

As a member of the defensive team of a B-52D crew, I actually had little reason to have physical contact with the weapons carried in the belly of the big bomber. My job as the Electronic Warfare Officer, like the gunner's job in the tail, was to insure that we successfully penetrated the enemy's defensive systems to get our aircraft into the proper place to deliver the weapons. I had radar jammers, chaff, and flares to help me to perform my task. The gunner had a more invasive weapon with his quad 50-caliber machine guns. Each of us knew what we were expected to do and the importance of our jobs to protect the rest of the crew during a wartime mission. I also held the combination to the mission material box that both the pilot and I would have to unlock to get to the tickets that held the go-codes that would send us on the way to drop the bombs. My only real active involvement with the weapon delivery system was my requirement to pull the special weapons manual lock handle that was at my crew position's feet. Without me pulling the handle and locking it into position, the radar navigator would not be able to release the weapon onto the target. It was part of the two-man policy and protection system built into the control of nuclear weapon delivery that made it impossible for one person alone to drop the bomb.

The pilot and copilot of the flight crew in the front cockpit had similar relationships to the actual nuclear weapons. Their jobs were to fly the plane and manage the flight resources that got us over the target. They had to get us airborne, perform the in-flight refueling, and negotiate the terrain avoidance and pop-ups necessary to get the weapon to the target. They, like me and my gunner, had no reason to come into physical contact with the weapon.

The two members of the nav team downstairs were the ones most associated with the actual bombs. Granted the navigator was the one who actually guided the aircraft to the target area, and once there the radar navigator took over and did the aiming and release of the weapons. In reality, the navigator and especially the radar navigator really had a reason to actually touch the weapons during preflight. With the two-man policy requirements, the radar navigator could not perform the preflight solo. For that reason, the navigator and sometimes others of us on the crew often accompanied him into the bomb bay during the assumption of alert pre-flight. It was then that the bombs' settings and serial numbers were checked and cross-checked and verified.

Let me make it perfectly clear that I can neither confirm nor deny the existence of any nuclear weapons aboard any aircraft on any base where I was stationed. I therefore cannot say where the weapon that I touched was actually located, but I will say without a doubt that I actually touched one. Trust me.

When I entered the bomb bay and saw it hanging in the rack, I was set back in awe. I remembered building a plastic model of a B-58 Hustler aircraft when I was about 10 years old. It had a plastic button on top of the model and when you pushed down on it, the plane's bomb bay door opened and a silver plastic A-bomb would fall out of the aircraft. The plastic bomb of the tiny model was a sleek looking aerodynamic bullet-shaped device – nothing like the object I was looking at in the real bomb bay of the real B-52. It was almost 10,000 pounds of death and reminded me of a propane tank in its size.

Now I suppose that if I had really done my mission planning in a true professional manner, I would have known the nomenclature for and what to expect the weapon to look like before I ever saw it. But I didn't. I had other things to worry about. At the time my wing commander was one Col David E. Blaze, a man who scared me to death. No man on earth ever reminded me more of Frank Lovejoy's portrayal of a Curtis LeMay character in *"Bombers B-52"* than Colonel Blaze did. He demanded that the EW officers in his wing have extensive facts memorized about the threats his aircraft might encounter. He expected me to know the NATO names of the fighters, the names of the radars aboard them, the number of missiles each carried, and the range of the missiles. I didn't have time or the mental capacity to worry about the characteristics of the weapons we carried.

I knew we had multiple targets and multiple weapons and I guess I heard the radar navigator brief the weapons' specifications to the wing commander when we were certified for our crew's Emergency War Order mission. I know I heard megaton this and megaton yield that and ground burst and air burst, but to my EW brain, it didn't matter. It did not matter as much as surface-to-air missile defenses, SAM ranges and altitudes and air-to-air weapons carried on Floggers, Flaggons, and Foxbats. The MK-53 designator of our own bomb was not as important to me as the burn-through range of the Fan Song radar.

But that first day I walked into the bomb bay and saw it, I knew that I had to reach out and touch it. I took off my Nomex glove and cautiously and deliberately pressed my bare flesh against the metal casing of the bomb. I held it there, almost as if I could absorb some of the power from it like a hero in a comic book. It was not enough just for me to look at it; I had to touch it.

If ever there was a single instrument that taunted the idea of death from above, it was the nuclear weapon carried aboard the B-52. No other apparatus ever created by the hands of man has ever had the potential for the death and destruction such a weapon possessed. It beckoned me like a wet paint sign on a shiny park beach. I reached out my hand and laid it on the cold, hard, metal object. I don't know what I expected to feel or what I expected it to feel like, but I knew that I had to touch it. I had to be able to someday say I had touched it.

And now, both the B-52D and I have been retired for many years and are in our own versions of a boneyard of one type or another. Many aircraft and former crew members cease to exist entirely, their memories the only proof of their efforts. Carswell AFB, Texas, as I knew it, no longer exists since it was effectively given to another branch of the service. Even the mighty Strategic Air Command is no more, having fought and won the Cold War task for which it was created.

In my civilian occupation I often find myself looking around at the objects that fill my work day and office space. Even though I work at a major university and am surrounded by great minds and great thinkers, I sometimes laugh at the panic my co-workers feel when faced with what they consider a major crisis in their jobs. Crisis? They don't even know the meaning of the word. Add the word "world" in front of "crisis" and then you approach the challenge we faced as Crewdogs. Even though I am involved in technology and computer science, there is nothing that I work with or do in my job today that could ever hold a candle to the awesome technology and responsibility with which I dealt in the Air Force over 30 years ago. My good friend and former fellow Carswell Crewdog, Vince Osborne, once told me this: "I have frequently said to someone, 'I'm not going to worry about this; I've been shot at by professionals.' Sometimes the statement gets a little chuckle." He added that the people don't really grasp the idea that the statement is true but that it helps him get things into perspective real fast.

Nothing could ever compare to the feeling I had that moment I stepped into the bomb bay for the first time and saw and touched a nuclear weapon. Civilians don't understand. They could never understand. There is nothing that I do today that makes me feel as important or gives me the sense of service to my country that I had all those years ago as a member of the defensive systems team of a mission ready combat crew of a B-52 Stratofortress. Civilians can't comprehend what it took to be a Crewdog, both physically and mentally, during the Cold War. How could they? Unlike me and my fellow Crewdogs, "peace" was never their profession. They've never actually touched an atomic bomb or never actually planned for or flew a B-52 mission that threatened and sometimes delivered "Death From Above."

The author and crew E-21.

Last Alert Tour

Bill Dettmer

Last flights in an aircraft usually mark passages from one phase of a person's life to another. In the former Tactical Air Command, such occasions were often commemorated with the aid of the flight line fire department, whose members were eager to interrupt the many hours of duty boredom with the occasional "hose-down" of a fighter pilot flying his last mission.

In SAC, such hose-downs happened occasionally, but having six members on each B-52 crew coupled with the turbulence of substitutability and reassignment made flight line wet-downs impractical most of the time. Not so with the last alert tour, however.

Because only integral crews ever sat alert together, and because alert schedules were pretty well fixed for a few months in advance, when a crew member was notified of a pending permanent change of station (PCS), permanent change of assignment (PCA), or separation, the time of the final alert tour was usually well known in advance.

I remember the first wet-down I ever witnessed. I was a navigator at the time (though I later went to pilot training and returned to B-52s).

As my crew, E-16, completed an alert tour in mid-summer of 1969, the six of us shoehorned ourselves into our sortie's assigned "six-pack" crew-cab pickup, and the oncoming crew gunner drove us from the alert aircraft toward the alert pad gate. I sat in the back, jammed in with our radar nav, EW officer, and gunner. Up front, in the middle beside the oncoming crew's gunner, sat Captain Billy Tedford, our copilot. In the "shotgun" position sat our crew commander, Tim Westover.

Just before we reached the gate, Tim said to the driver, "Gunner, would you please detour past the alert facility swimming pool? We have somebody that needs to go for a swim." Like many a SAC copilot, Tedford had received a Palace Cobra assignment—a one-year tour to Southeast Asia. This was his last alert tour before reassignment.

At Beale Air Force Base, the "Rock Hudson Memorial Swimming Pool" lay within the secure perimeter. The gunner driver took a sharp right and pulled up beside the pool, which sat on a small rise above road level. Tedford knew he was trapped and couldn't escape, but he figured he didn't have to cooperate. All of us in the back seat bailed out and surrounded the front passenger door. Tim Westover got out, and we all grabbed the copilot, dragging him forcibly from the six-pack, in spite of his vain attempts to hang onto any part of the vehicle he could. With one of us on each leg and arm, and the fifth holding up his butt, we schlepped Billy to the edge of the pool and heaved him in.

This process became a semi-regular event at the Beale alert facility over the years, so regularly practiced that crewmembers came to expect it on their last alert tours. Sometimes individuals would escape the ritual dunking when unexpected developments would result in someone having had a last alert without realizing that it would actually be their last. But this happened infrequently, at best.

Fast-forward seven years to early 1976. I had left SAC in 1971 (after my own wet-down) headed for pilot training. A year later, by personal choice (and not a small amount of manipulation of the personnel system), I returned to Beale AFB again as a B-52 copilot. Some three years later, I was commander of my own crew, E-21, composed of copilot Captain Bill Lovell, radar navigator Captain Jack Wright, navigator Captain Jim Fitch, EW officer Captain Arnaldo Hernandez, and gunner Sergeant Jim Ryan.

Arnie Hernandez hadn't been on our crew very long, but it was long enough to develop the bonds of both friendship and professional respect. Arnie was quiet but friendly. A graduate of the University of Puerto Rico in electrical engineering, he was commissioned through Air Force ROTC and assigned to navigator training at Mather Air Force Base. His electrical engineering degree made the Electronic Warfare Training School a natural progression. The 456th Bomb Wing (later redesignated the 17th Bomb Wing) at Beale was his first operational assignment following EW training.

And it turned out to be his last. While Arnie and his lovely wife Ana undoubtedly enjoyed the opportunity for travel and new experiences that Air Force active duty provided, they longed to return home to Puerto Rico. By February 1976, Arnie's active duty commitment was drawing to a close, and he had submitted his separation paperwork. Not coincidentally, this process also established the dates of his last alert tour.

Unlike many departing crewmembers who accepted the prospect of a final alert wet-down stoically, Arnie was determined to avoid it. Apparently he had expressed this determination to Jim Fitch, our navigator—who, naturally being the loyal crewmember he was, immediately advised me about it! I decided we needed to do something special for Arnie's last alert tour.

Because the toss into the Rock Hudson Memorial Pool after alert changeover had lost its novelty, over the years crews had invented new schemes for getting their departing members into the pool. In most cases, these schemes evolved into "doing the baptism" by surprise a day or two before the final alert tour actually came to a conclusion.

The fact that this was to be Arnie's last alert tour was no secret. Arnie knew it. We knew it. And most of the members of the other crews on alert knew it too. However, nobody mentioned it. Arnie reported for that alert tour with no small amount of trepidation. He knew he would likely be ambushed sometime during the next seven days, but he didn't know when. As a matter of fact, neither did we. We hadn't figured out exactly how we would complete Arnie's "baptism." So for the first four days of the alert tour, we let Arnie's suspense build up, acting as if we had forgotten all about it. But we hadn't, and Arnie knew we hadn't.

As each day passed without a dousing, Arnie's tension grew. But he also knew that the window of opportunity was shrinking, and whatever he could do to close it some more on his own improved his chances of getting away dry.

By Wednesday, our sixth day of alert, Arnie was in full avoidance mode. That afternoon he spent an additional two hours over in the squadron after his T-4 trainer was done, just to keep from exposing himself to the rest of us back at the alert facility. When the squadron staff finally shut down for the day at 1630, Arnie drove the crew vehicle back to the alert facility, but he didn't make an appearance upstairs. Instead, he went directly down to his room, carrying food and snacks he'd obtained from the Base Exchange annex in the big composite building the wing operated from. He planned to eat in his room and go to bed early.

Jim Fitch found out about this tactic and mentioned it to me while the rest of us sat down for dinner in the alert force dining room. Then we strategized.

We dispatched an EW officer from another crew to Arnie's room to see if he could entice Arnie out. But not only had Arnie retreated to his bedroom, he'd fortified himself in it! He'd moved the ranch oak dresser in front of the door and backed that up with desk, chair, and wardrobe. Even though the doors had no locks, it would be virtually impossible to force this one open from the outside. We had to figure out a way to get Arnie out of the room voluntarily.

There was a certain issue of creativity involved here, too. Perhaps we could lure Arnie out of his room, but just to throw him in the pool? That wasn't too original. I mean, crews had only been doing that for eight years. An effort of this magnitude demanded something new, something special. Bit by bit, the idea fell into place as we sat around the dinner table.

We decided to wait a few hours, to put Arnie off his guard while we organized the plan. In fact, we waited until nearly 2300 hours. Arnie had actually gone to sleep by this time, his ranch oak barricade securing our only access to him.

Around 2030, I convened all the bomber and maintenance crews in the briefing room for a "mission planning session." With four full bomber crews (minus one person) plus crew chiefs and their assistants, we had a formidable strike force—one experiencing the usual seventh-day boredom of repetitive alert tours and eager for the distraction of novelty.

"The mission," I told them, "is to thoroughly wet down Arnie Hernandez. This mission is complicated by the fact that he's not only an unwilling participant; he's also barricaded himself in his room. We have to figure out a way to get him to come out voluntarily."

We had one thing going for us that Arnie could not ignore: duty (as in "duty, honor, country"). We all were out there in the first place to respond to our aircraft if alerted to do so. All we had to do was stage an alert, and then douse him somehow. But that in itself posed another problem.

Neither SAC nor its combat security police had much of a sense of humor where nuclear weapons were concerned—certainly not a bad thing for national security, but it did put a bit of a crimp in our plans. None of us wanted to risk our careers on a practical joke. Staging a fake alert inside a nuclear facility without the knowledge of the command post would be bad enough, but we had sky cops with loaded weapons out there. The potential for things to get out of hand was substantial. But we decided to deal with this obstacle after the concept plan was fully formulated.

I thought about the architecture of the alert facility. If we actually did simulate an alert, and fooled Arnie into believing it, he would come out of his room and head out one particular set of doors leading to the ramp running upward half a level to the parked crew vehicles. If we could get him into that half-underground ramp, we could douse him from above. But how to irrigate him?

There was one long garden hose outside used by the transportation squadron to wash the crew vehicles while they were parked on alert, but that wouldn't do. Not enough "bandwidth" for sufficient water and certainly not dramatic enough. Somebody suggested the G.I. waste paper baskets scattered around the alert facility. You know the ones I'm talking about—the ubiquitous, gray five-gallon waste cans that every room in every Air Force facility ever ran. A couple of people ran off to inventory the trash cans in the alert facility and reported back a few minutes later: a total of 28. Our hydraulic capacity was 140 gallons.

When the time came to execute the plan, we would collect all the cans and line them up along both sides of the wall enclosing the ramp. (Unlike the ramps at northern-tier bases, ours needed no snow shield, so it was open to the sky above from the moment one passed through the double-doors from the sleeping quarters below.) The hose would be used to fill these cans, 14 on each side of the ramp, about six feet up. It turned out that this required the better part of half an hour.

The plan was to have the entire alert force out there in the dark manning the buckets. They were to hold until Arnie was well out the door and onto the ramp. But what if he happened to glance up and saw the reception committee? He might turn around and dash back inside, and our element of surprise would be lost. We had to make sure he was comfortable dashing out the door, yet be able to secure it behind him.

Our copilot, Bill Lovell, would be detailed to bang on Arnie's door and tell him we had a klaxon-out alert. Bill would keep yelling at Arnie just to boost his anxiety as he tried to clear the furniture away from the door. Then, when Arnie hit the hallway, Bill would take off heading for the double-doors leading to the "canyon of death." He'd blast through the doors and up the ramp. The dousing team was briefed to let the first guy pass and hit the second guy.

While Arnie was pursuing the sprinting copilot, Jim Fitch would fall in line behind him, egging him to hurry up. As soon as Arnie cleared the outer doors, Jim was assigned to pull them closed from the inside and hold tight to the panic bars, so Arnie couldn't retreat to the safety of the building. The plan was shaping up!

All that remained was to clear it with "the authorities." No, not the wing commander or director or operations—if you can't stand the answer, you never ask the question. In this case, at 11 o'clock at night, it was the command post and the combat security center (CSC) we were concerned about.

One of the benefits of being the senior aircraft commander on alert is that during off-duty hours, I was basically in charge at the alert facility. Once the plan was finalized, I briefed the alert force charge-of-quarters (CQ), a two-stripe airman, on what we were going to do. Then, in his presence, I called first the command post and then the CSC.

The command post was easy. The controller was a former squadron member. I told him what our objective was and that we were going to simulate a klaxon-out alert. I said if anybody reported anything unusual at the alert facility, he shouldn't worry—and most important, not report anything to higher headquarters! He was completely on board with what we were going to do. I then coordinated with the supervising NCO at the CSC, explained the plan to him in detail, and suggested he verify what I'd said to him with the command post, which he did. I asked him to brief all perimeter and aircraft guards that we'd be starting the crew six-pack engines and turning on their rotating beacons, but that they would not be moving from their assigned parking spaces. This was a way of making it seem more real to Arnie, because one or two of the vehicles could be seen from the bottom of the ramp. I concluded the calls by telling them the precise time we were going to execute this: 2305 hours.

At 2300 the alert facility was completely empty, except for the alert force CQ, Arnie, Bill Lovell, Jim Fitch, and of course me. (I had elected to watch the proceedings from inside the double doors with Jim Fitch, and help him hold the doors closed if required.) Everybody else was out at the "water combat line." A couple of the gunners started all the six-pack engines and turned on their beacons. Fitch hid out in his bedroom, with the door slightly ajar; waiting for Arnie to pass by, and ready to push him out through the outer doors should he smell a rat.

The alert force CQ followed me downstairs with his bullhorn, went to the hallway intersection closest to Arnie's room, and announced: "Crews report to your aircraft, do not start engines!" He repeated this several times. Lovell pounded on Arnie's door and tried to open it—unsuccessfully, of course—yelling, "Arnie, come on! We've got an alert! Hurry up!"

From within, we could hear Arnie stumbling out of bed in the dark, trying to get his flight suit on, and desperately moving his barricade away from the door. I backed off around the corner, where he wouldn't see me. Finally, the door opened, and Arnie stumbled out, eyes wide, barefoot, carrying his big black navigator bag and his flight boots in both arms.

"Come on, Arnie!" Lovell yelled again and sprinted for the doors to the ramp. He hit the inner doors (there were two sets, with a small weather vestibule between), blasted through the outer doors, and was at the top of the ramp looking back in a heartbeat. He was up that ramp so fast that if the water crew had accidentally dropped early on the first guy instead of the second, they'd have missed him!

Arnie pushed into the weather vestibule and started opening the outer doors. He was part way out when he happened to look up and saw the bucket brigade and stopped short. No such luck, Arnie. Fitch shoved him out and pulled the doors closed. He and I held the panic bars to ensure Arnie couldn't get them open from the outside. With a series of war cries, the buckets were tipped, and H_2O flowed.

From below and inside the outer doors, the sight was truly impressive. Picture the scene from *"The Ten Commandments"* when the Egyptian army gets drowned: Arnie was like an Egyptian charioteer as the Red Sea closed in on him—amid no small amount of hooting and hollering from above. I learned that 140 gallons of water in that confined space actually creates a four-foot wave. Like a tsunami, the wave washed over Arnie and smashed against the outer doors, rolling all the way up to the windows before rebounding. Arnie got it from both directions. Very little of the water seeped in below the molding of the weather doors, barely enough to wet the vestibule floor. Arnie stood forlornly looking in at us through the glass, still barefoot in water almost up to his knees, saturated up to his armpits, and clutching his boots, socks and nav bag. Almost reluctantly, he smiled a crooked little smile: "Okay, you got me."

The now "clean" waste baskets found their way back to where they lived. I called the command post and CSC to report "exercise complete" and a battle damage assessment. Arnie went back to take a shower and go back to bed, but I don't think he rested easy the remainder of the night.

Of such adventures was alert duty made. "Sleep tonight, America, your Air Force is awake." Well, some of it, anyway.

Planning for the Unplanned

Rich Vande Vorde

The B-52 training mission lasted eight to 10 hours and the flights were packed full of activities for us to perform. The aircrews were scheduled to plan and coordinate the missions the day before. I'll share with you the Navigators (Nav) perspective on one particular mission.

As the Nav, preparing for a mission started before the rest of the crew. For a flight on a Monday, the mission planning began the Friday beforehand. When the flight schedule came out on Thursday near noon, I would take the information and start working. I would plot the activities on the Navigational Chart and prepare a Form 200 with specific times and tracks in order to get to the appropriate activities on the control times. Note: this was back in the days before fancy computers generated specific missions to plug into the navigation and bombing system. I was manually computing back then.

This specific mission was very snug on timing. Our mission was from Carswell AFB, Texas, to Richmond, Kentucky, flying a low level route with air refueling in between. So, during mission planning I, as the Nav, emphasized this fact and we discussed alternatives just in case the mission didn't go as planned.

Everyone on the crew understood that completing the low level portion of the mission was the most important activity. Most of us on our crew needed to complete a low level that day in order to stay current and therefore to be able to pull Alert the next week. We had an extra pilot on board with us that day as well. With all that in mind, we discussed the latest takeoff (TO) time for the following: going direct to refueling, going direct to the Richmond Primary Entry Control Point (PECP), and going direct to the Richmond Alternate Entry Control Point.

When we arrived on Monday for the flight, there was a problem with the aircraft. We learned early in our training that if something could go wrong, it probably would. So we learned to expect last minute changes. We planned for the unplanned. Thank goodness for that full day of planning beforehand.

We had to wait at Base Operations until the aircraft was ready, so we gathered in the Planning Room to discuss the impact of the delay that we had previously discussed on mission planning day. Could we drop two of our three scheduled bomb runs and still get to the entry point at Richmond to meet our last bomb run control time? I calculated that we could take off two hours late and still get there in time.

Then maintenance personnel notified us that the aircraft was ready. I emphasized to the crew that if we were swift with our preflight, we could actually takeoff only about one hour late. That meant that Air Refueling (AR) was back on, but there would be a delay.

The preflight and takeoff were smooth and it was a quick trip to AR. The Instructor Pilot (IP) was pleased that we were able to complete it. But as we exited the AR track, the pilot noticed lightning and dark clouds in the distance. So the Radar Nav was asked to check it out. About 120 nautical miles directly ahead was a line of thunderstorm weather radar returns perpendicular to our course. The RN directed the pilot to turn northeast toward a break in the line. Maneuvering was closely coordinated with the Air Traffic Control (ATC) center. The tricky part was to stay at least 20 miles from the line of thunderstorms and still get to Richmond on time. I was continuously checking and advising the crew whether we could get to low level for any activity. We finally cleared the last weather hazard and headed PECP direct. I notified the pilot that we needed 500 knots of airspeed (nearly maximum), in order to get there on time for our last scheduled release time. After ATC center approved our new airspeed and track, the crew completed all the required checklists to prepare the aircraft for low level.

We were about 100 NM from the PECP when the radio conversations became very intense. With the 80-knot tailwind we were traveling about 550 knots ground speed, far faster than normal, which made these activities even more intense. I remember two or three ATC areas intersected in the vicinity of the PECP, as well as a commercial airway. It was a good thing we had an extra pilot on board to keep track of frequencies and actions required by multiple controlling agencies. The entire crew normally kept track of this information too. We had to be cleared to enter, maneuver the aircraft to the correct heading, and start a descent at the appropriate time and position.

Needless to say, the RN and I were working just as hard to get good information to the Pilot controlling the aircraft. We had to allow someone to talk to center for clearances without everyone talking at once.

Well, I'm not sure if we were really cleared to enter Richmond that day, but I do know we really did appreciate having thought about the options during our mission planning day. It helped prevent errors that could have been disastrous. We were ready for the unexpected because we planned for the unplanned.

This is probably a familiar situation for anyone who has flown from Carswell to Richmond. However, after we landed, the Squadron Commander and Operations Officer met us with a question for our crew. Had we been "cleared" to enter Richmond? The IP's response was, "With the countless radio conversations, I am certain that one of them 'cleared' us to enter Richmond!"

Another Fun Time In the North Country

Richard Gaines

This will be a short story of winter flying in North Dakota. It is short because the incident lasted just a few minutes. Most flying experiences that shift your heart rate into overdrive last only a few minutes and sometimes only a second or two. Flying out of Grand Forks AFB, North Dakota, in the winter led to quite a few such occurrences. To digress a little, I will tell you a little of how bad the weather was back in the 1960's at Grand Forks AFB. A short time before Bunny, Lee, and I arrived in January of 1968, the worst snow storm in their history had dropped about six feet of snow in two days adding to the two feet or so on the ground that had already accumulated. The firehouse and every other single story building were unrecognizable as such. No private vehicles could move for five days and the sick and pregnant were taken to the hospital by Sno-cat. The temperatures were also unnaturally cold in the late Sixties. My first four flights in the B-52 were in takeoff temperatures of around minus 30 degrees. That minus 30 became very common for the next five winters we were there. Looking back at historical meteorological data for those years at Grand Forks, I found that six all time low temperatures were set in that time. Seems like a "banana belt" up there now, according to the weather channel.

Back to this story. We were ending a 10-hour mission from GFAFB and taking heading vectors from Grand Forks Approach Control. The current weather at "The Forks" was two miles visibility and a cloud ceiling of 500 feet. The wind was 150 degrees, gusting to 45 knots. That was going to be a bit sporty but with the B-52's strange cross-wind landing gear it was manageable. I was vectored around the north side of the field and started an ILS approach at about seven miles. At three miles we could see the lights from that beautiful 12,000' SAC runway fading in and out. At a mile from touchdown, the runway was visible. We crossed the end of the runway and I flared the airplane to touch down on the aft landing gear. At about ten feet and the throttles at idle, we were hit by a complete "Whiteout" of blowing snow.

Everything we could see a second before was gone. No runway, no runway lights and no lights from the buildings. I immediately came to full power but not before we landed on something which I presume was the runway. The eight big fan jets immediately came up to full power and we were flying again in a split second. That little maneuver took everything I had in flying skill. The airbrakes were full up for landing and the B-52 was trimmed for a power off touchdown. I had to use full forward control column as I re-trimmed and dropped the airbrakes carefully back to off. We flew runway heading back to 1500' and got the flaps up and informed the Command Post that were going to K.I. Sawyer AFB, a couple of hundred miles east. The enlisted ground crewman that was along for the ride was scared out of his wits but recovered rapidly. We flew back to home base the next day in beautiful, cold weather.

Chapter Three

Inflight Emergency [in-flahyt] [i-mur-juhn-see] – *noun* – A state, esp. of need for help or relief, created by some unexpected event done, served, or shown during an air voyage.

Close Call

Peter Seberger

In the spring of 1968 there was a substantial shakeup in at least my part of the SAC bomber force alignment. This was due in large part to the extensive involvement of the D-model fleet in the Arc Light bombing campaign. I had been stationed at Carswell since June of 1965 and my crew had participated in the second through the fifth along with 22 other Arc Light missions between July and December of 1965. At that time the Columbus and Mather units relieved us. In early 1966 the big belly modifications made the D-model fleet more effective, and they assumed the primary Arc Light role.

Since the D-model units pulled so much TDY, the F, G, and H-model units, as well as the few remaining B-model units assumed the primary stateside alert role. The inception of the RTU training program began sometime in 1968-69. This RTU program gave some relief to the D-model unit crew forces by cycling G and H-model crews through a two-week differences course at Castle, followed by a five to six month TDY. That allowed them to enter a crew rotation schedule at Guam in 1969 for initial training, followed by Kadena and U-Tapao rotations. The 30 F-models at Carswell were sent elsewhere, and replaced by 15 D-models. About half of the bomber crew personnel at Carswell were transferred out, mostly to the Northern tier bases.

As others have related, if one did not yet have a Northern tier assignment on his record, he was almost invariably sent to one of them. I was sent to Grand Forks AFB to fly the H-model, and was ordered TDY enroute to attend a two-week G/H differences school at Castle AFB. I had a new son, and so I left my wife, son, and Irish setter in Nebraska with family before driving out to Castle. The school was scheduled for two weeks, and included considerable classroom and three simulator missions, concentrating on the aircraft differences. As I recall, Castle did not have any H-model simulators available at that time so we used G-models for all our training.

When the training was finished I made a quick trip back to Nebraska with another copilot from Carswell who had attended the same course. I put him on a bus to finish his trip wherever he was going, and spent a few days leave in Nebraska before taking my family up to Grand Forks. When we arrived there in early May, I signed in to the squadron and registered for base housing. Because Grand Forks AFB was also home to a large missile unit, the personnel situation was always active, and it turned out there would be about a three-week wait before we could get base housing. The only option for us was a motel about ten miles east of the base which was close to the airport and city of Grand Forks. It wasn't fancy, but they were accustomed to transient military clients and had a rudimentary kitchen which we could use, so it filled the square for the short time we expected to be there. It was pretty crowded with a four-month-old baby and a 100-pound dog but not as bad as it could have been, so we made the best of it. The dog was unfortunate enough to get his basic skunk survival training during that time.

In the meanwhile, the 46th Bomb Squadron needed aircraft commanders and I was a relatively high time copilot. I had actually flown two or three upgrade missions at Carswell, but was still fifth down on the Carswell list of upgrade eligible copilots when I transferred. One of my pilot training classmates, Don Lewis, had been sent to Grand Forks for his first assignment and had already been an A/C for a year when I arrived. I was eager to fly and upgrade, since I already had nearly 1,500 hours in the B-52 and had not been happy with my upgrade progress at Carswell. So after a few days of indoctrination and ground training I was scheduled for a flight. It was not an upgrade flight but was a chance to go along and see how things were done there and perhaps get a bit of stick time. We would do whatever turned out to be convenient given the needs of the primary crew. I figured to get at least a couple of approaches and landings, but probably not much else. It was not exactly a dollar flight, but it was a chance to fly, so I was glad to go.

I don't remember much about the mission planning, but I do remember that the instructor was the assistant DCO, Lt Colonel Al Selander, who of course had many other duties. I don't think he attended any of the primary briefings though to be honest I can't remember that part of it. The A/C for the crew was Ernie Giepel, an experienced pilot on the base. Likewise, I don't remember the takeoff time or anything much about the mission until about 100 miles out from the base. They didn't start recording tail numbers on my Form 5 flight records until August of 1969, but I remember the tail number that day was 1031, (61-031). That same record shows me credited on May

17 with 7.7 hours total time, the penetration, three precision approaches, one non-precision, and one touch and go. That is the reason for this story.

I had spent some time in the seat during the mission prior to my touch and go, but for whatever reason I seem to remember getting back into the right seat about 100 miles prior to the penetration fix. That was what we used to do mostly, before the enroute approach became common to save fuel and time. Colonel Selander got into the left seat sometime later and must have been there through the approaches, though again I don't remember any of them. I do remember the touch and go, but not any specific briefing from the IP on final prior to landing.

My records show that my previous landing had been a nighttime full stop at Carswell, logged March 4th. So it had been some 75 days, a PCS, TDY, leave, school, and a different model aircraft to boot since my last landing. My A/C at Carswell was not an IP, and unless we had an instructor aboard, which seldom happened, there was only one landing per flight. We generally alternated that unless weather, broken airplane, or landing with nuclear weapons (like from a Chrome Dome) was involved. I did get my share of air refueling, and was not a bad stick from the right seat, but my A/C seldom practiced for more than a few minutes, preferring to get the fuel and drop away. When I was doing my initial training at Castle I had been on a two copilot student crew without an A/C, so we each got a chance to do our share of refueling and I could do a credible job when I left Castle in 1964. I remained on the same crew for nearly three years at Carswell, though, so I did not get much chance to practice there.

I thought the approach was good, and I flared to what I anticipated to be a normal landing. The touchdown was good, and the checklist was, as I recall, Airbrakes-SIX, Stab trim-RESET, Airbrakes-OFF, Power-STANDUP. I do not recall anybody making the point that this was the critical part of an H-model touch-and-go during my training at Castle. Neither was that point made during the mission briefing or prior to the landing. The TF-33 was one of, if not the first, turbofan engines. It was really mostly the guts of a J-57 remodeled to include some larger fan sections and an additional turbine to drive them. It had/has some funny characteristics during spool-up which could and generally did include a tendency to compressor stall when mishandled. There was a surge bleed valve which, if it worked properly, was supposed to take care of that problem, but it was an imperfect fix.

Anyway, my prior experience was with F-models and I was too aggressive with the throttles. I advanced them too far and remember the airplane drifting to the left. I thought I applied sufficient rudder to correct it, and when that didn't work and there was no input from the IP, I decided to go ahead and fly. The plane was trying to lift off, and was already well left of centerline. Anyway, I must have advanced the throttles through the thrust gate and at about the same time the IP reached over behind my throttle hand and tried to restrict the two outboard throttles on the right side. I found that out later during the debriefing and subsequent investigation. I might have canted the throttles a bit sideways inadvertently, but I think the right side throttles were misrigged, according to the maintenance folks later. This condition brought all the right engines up to power earlier than the left ones. Anyway my throttle handling, along with the restricted throttles on the right side, along with the misrigging if it was there, along with the fact that we were now at flying speed, all resulted in the left engines spooling up and putting out good thrust and the left wing flying immediately. Unfortunately, at the same time the right inboards engines started compressor stalls because of the mishandling and sideslip. The outboards were still retarded but also started compressor stalls according to witnesses on the flight line.

Needless to say, the right wing was not flying well so it went down as the left wing came up and the right tip gear and the fiberglass fairing on the bottom outside edge of the right wing kept the wing from digging into the earth and sending the plane cart wheeling. About that same time the IP gave the "I'VE GOT IT" command as he retarded all the left engines to idle to prevent that wing from rolling the airplane. He later said it was the hardest thing he ever had to do. The aircraft heading had changed alarmingly during this maneuver, of course, and some of the tower personnel later admitted that they hit the crash button. My throttle push had given us enough flying speed and along with ground effect (and undoubtedly significant help from the Almighty) the IP was able to get the right engines on line and the left ones back to generating thrust so we could maintain wings level. By then we were heading pretty much east instead of the runway heading of north.

The IP flew out and got the plane back heading pretty near north and then shakily said, "You've got it back." I remember looking at him and he was nearly white. I probably was too, as I knew the trouble we had been in. Then the Radar Nav said "Would anybody care to explain that last maneuver?" That broke the tension, and the IP suggested that Ernie get into the right seat. On downwind he did that and we made a low pass so that the Supervisor of Flying (SOF), who that day was my old pilot training classmate, Don Lewis, could look at the right wing and tip gear through binoculars. He informed us that the wheel didn't look right, so we transferred about 5,000 pounds of fuel to the left wing to give a 10,000 pound imbalance, and landed straight ahead on runway 35. We stopped on the runway and the maintenance guys came out, changed the wheel and hub, and we taxied to parking.

An inspection showed that there had been a completely destroyed wheel, hub, and tire on the right tip gear. There was also a nice scrape about two inches wide and several feet long where the fiberglass tip of the right wing had contacted the runway while the tip gear wheel and tire were destroying themselves. I saw that the wheel and the bearing housing in the hub were loose and all the cast aluminum spokes between the hub and rim on both sides of the wheel were cracked or broken. The tire was completely shredded. But the strut had held and every day since I thank God for the fact that whoever put that assembly together did his work well.

Had Colonel Selander hesitated any longer or anything gone differently the tip would have dug in and we would probably all have been toast. I went out later to the runway where we touched and looked at the marks made by the wheel, and could not believe the tight radius of turn. There was a significant gash in the surface of the runway, which I can still see today when I close my eyes. It is funny how some things stay with you. Surprisingly, due to the fact that the replacement parts value fell under the reporting threshold, the whole episode was not even classified as a flying incident, much less an accident and to my knowledge the wing did not report it further. Later on there was a complaint published in the base newspaper from a resident of the base housing about low flying airplanes over her house, but I never saw it.

However, the wing still needed bodies so after due reflection and conferencing, and not a little soul searching by myself as well as some others, I suspect, I was put on a crew with an IP named Jim Allen. Three days later he and I and a single navigator to run the heading and radar systems went out for some intense pattern work. My records show that on 20 May I logged 13 touch and go landings, two full stop right seat landings, four approaches and a penetration. I seem to remember that Jim made several landings himself demonstrating different throttle techniques, and after that flight I never had trouble landing again. Well, not until my initial A/C check ride, when I landed fast, got into a porpoise maneuver and the evaluator took issue with my recovery technique. I subsequently entered the upgrade program and went on to log pilot time in the BUFF every year after that until I retired with 3,100 BUFF instructor hours and 6,600 total BUFF time.

After I became an instructor I told this story to every pilot with whom I flew who was due to make a landing. The moral of the story: never hurry mission planning, know what is expected of you, be sure everybody else does too, and handle those throttles carefully!

The author.

My Perfect Three-Point BUFF Landing

Bill Jackson

The summer of 1966 was a busy one at Glasgow AFB, Montana—that jewel of a SAC installation. Located 60 miles south of the Canadian boarder and 45 miles north of the Fort Peck Dam, it was the home of the 91st Bomb Wing complete with the 322nd Bomb Squadron, the 907th Air Refueling Squadron, and the necessary supporting units. Keeping them company was the 13th Fighter Interceptor Squadron.

It was the northern most Air Force installation in the continental limits of the USA with the exception of an ADC radar site located a few miles up the road from the base. The wing had just been notified that it had won an outstanding unit citation award for the previous year and was more than busy preparing for the coming Arc Light deployment to Guam in September. The fact that it was 296 miles to Great Falls, 290 miles to Billings, and seven microwave antennas transmitting the only television signal from Salt Lake City made the personnel stationed there feel sort of left out of things. Well, it was the only signal for over two years.

Every combat ready crew was required to make an annual minimum interval takeoff (MITO). On this particular day my B-52D crew was in line to be the number one aircraft for takeoff for our MITO training. We were to be followed by four other aircraft at short intervals. On this same day the bomb wing was host for an annual visit of several Air Force Academy cadets. Three of them were observer passengers on my plane.

We began our takeoff roll at the scheduled time. Everything appeared normal until just after liftoff when my tail gunner reported that something had blown off the aircraft. The gear had already retracted and I was about to start the flaps up. Everything was normal in the cockpit. There was no vibration so I notified the command post about what our gunner had seen. I informed them that I was continuing the mission as planned. The following four aircraft were safely airborne, all reporting seeing nothing unusual on the runway during takeoff.

At this point in the mission the most logical explanation of what our gunner had seen was that one of our tires had blown. After some discussion back and forth, I suggested that after refueling and the navigation leg that each aircraft should be cleared to drop a practice bomb. Each aircraft would have to descend to 10,000 feet altitude so the navigator could visually inspect the bomb bay doors. He could ensure that the bomb had actually departed the aircraft as required by a recent directive. At that time also the gunner could come forward from his tail compartment and inspect the aft gear tires.

The wing commander approved this suggestion, and the flight continued. The commander closed the runway, and hundreds of personnel were lined up elbow to elbow to walk the entire length of the 14,000 foot runway to look for the unidentified object. They found nothing.

After the bomb drop the descent was accomplished. The gunner reported that the tires appeared normal. The bomb doors were clear so a climb back to altitude was begun. An HF radio contact was established with the SAC command post, and a phone patch with Glasgow was made.

The Dash One stated that if a tire or tires had blown that the affected gear should not be extended. After much discussion I suggested that in as much as the five ships were still in cell formation that we lower the gear. Then the number two plane in the formation could visually inspect our aircraft for damage. SAC and our wing commander concurred. Number two was directed to close in and inspect our gear and the underside of our plane.

Major Atkins was the pilot, and he reported that the outboard tire of the right aft gear appeared to have a large hole in it. More discussion followed and it was decided that the gear should be retracted, that the mission should be terminated and that we return to Glasgow.

The bible (Dash One) had a paragraph which covered our situation: pull a circuit breaker, leave the right aft gear in the up position, lower the remaining three trucks, reduce the fuel onboard to 20,000 pounds which would be 10,000 pounds below recommended minimum for landing, make a normal approach, stay off the brakes and do not turn off the runway.

The weather was Ceiling And Visibility Unlimited (CAVU) but only about one hour of daylight remained when we arrived back in the local area. The gear had been lowered earlier to burn off the excess fuel. However about 10,000 or 15,000 pounds still had to be consumed before touchdown, so several low approaches were made.

All of us who could see outside the aircraft were amazed at the number of blue-suiters and vehicles that had gathered at the far end of the landing runway. A fire truck was at every intersection. The wing commander was concerned about the fuel remaining, so we had to give him a fuel reading every minute or so.

The EW had made sure that the three passengers were briefed on emergency procedures and that they were strapped in properly. As I began my turn on the final approach I announced on the interphone that every one should assume his crash landing position and that there was time enough, for those who wished, to do a quick tour around the beads. This remark was intended to put a little levity in the situation. But about 10 seconds later I felt a tap on my right shoulder. I turned my head and saw a string of beads being held out by a cadet passenger who was riding in the instructor pilot position. What could be seen above the cadet's oxygen mask was sincerity, and on interphone he stated, "Here sir, you can use mine." That is when I almost lost it!

I was able to recover in time to make a normal landing. The bird kind of sagged a bit on the right side, but otherwise everything was normal. The right rear tire was found to be the best tire on the aircraft! It showed no damage whatsoever. And what had blown from our aircraft during takeoff was never identified for certain.

The aircraft we were flying had not flown for over 30 days, so this was a First Sortie After Ground Alert (FSAGA.) Unusual malfunctions frequently occurred on FSAFAs. The best bet was that the paper placard proclaiming that high explosives were loaded in the bomb bay while the aircraft had been on alert had been overlooked and had not been removed by the weapons team.

Someone remarked that my gunner, Sergeant Armfield, had stated many times he would never live to see his retirement. It was also noted that this was his last flight before retirement.

So just another day passed in the lives of us Crewdogs. We experienced lots of hours of drudgery interrupted of a few moments of pure panic! But I must admit that three-point BUFF landings were not the usual order of the day. We did it once, and we made it look rather easy!

A Brush with a Near-Miss Disaster

Al Hodkinson

After completing a routine B-52D training mission my crew was returning to land at Fairchild AFB in Spokane, Washington. As we were approaching the base we called for landing instructions and were advised by Radar Approach Control (RAPCON) that Fairchild was closed because of a crashed aircraft on the runway. We were instructed to contact the command post for further instructions.

The command post requested an estimated time of arrival (ETA) and our current fuel status. After assuring that we had enough fuel to continue the flight, we were diverted down the coast to Travis AFB, California. Travis had two parallel runways and upon contacting RAPCON at Travis we were cleared to land east on the right runway.

We were on the base leg at 1,000 feet altitude, when the gunner reported that he had traffic at three o'clock level with us and approaching. Looking out to the right I saw the other aircraft was about to fly under our right wing! "I've got it!" I stated and applied full power to all eight engines and pulled back on the wheel.

We were very light and the BUFF responded and leaped upward. I glanced down again and watched the other plane go under us so close that I saw inside the cockpit and spotted the cabin lights.

I again asked the tower if there was other traffic and was told that there was none except for a civilian DC-8 landing on the left runway. I made a 270° turn and we descended and landed without further incident.

After debriefing we went to base operations and filed a near-miss report. The tower and RAPCON maintained that there had been no problem. I talked to the civilian crew of the DC-8, and they were totally unaware that a B-52 was attempting to land at the same time they were.

The other plane had 262 people onboard and it would have been a mess over the city of Fairfield that night except for our observant gunner. I have no idea what happened to the near-miss report that we submitted. I could not find any record of it later.

Our mission was completed a few days later with a flight in a KC-135 to Moses Lake and then by bus back to Fairchild AFB. The plane crash on the Fairchild runway was still the main topic of conversation when we returned home. I found out that it had not been a crash but was a B-52D which lost a wing while running up engines just prior to their takeoff roll

Thanks to our vigilant gunner we averted a major disaster which would have been of much greater significance than that of losing a wing on the end of the runway!

Arc Light 1966-67. The Crewdogs names are: Kneeling L-R: Capt. William A. "Shiney" Channell (RN), 1/Lt. William W. Redmond (N), SSgt. Paul A. Sholberg (G), Capt. James E. "Jim" Bradley (EW) Standing L-R: Maj. F. J. "Fritz" Joraskie (AC), 1/Lt. Garland "Gar" Pill (EW) who was stand-in for 1/Lt. James M. "Mac" Turner (CP) who was flying substitute CP on a mission. Note: "Mac" Turner was causality in a battle-damaged B-52D crash at U-Tapao during Linebacker II. Glasgow AFB was about 16 miles north of the town of Glasgow, Montana. When the 91st BW/322nd BS were notified of pending deployment to Guam on Arc Light the town's Chamber wanted to do something to give the crews a tie back home to Glasgow. So they purchased the Tams, one for each Crewdog. I think they were ordered from Scotland. Anyway they took the place of our yellow ball cap-style caps we wore back home in Montana. We never wore them on missions, just for photos. I still have mine here at home.

Command Post Checklists

Jim Bradley

My assignment to Glasgow AFB, Montana, was a unique experience. For one thing I was used to flat terrain instead of rolling hills and mountains. For another, we had B-52D models instead of other models. This made for interesting combinations of experiences such as those encountered during Arc Light and Chrome Dome flights.

Many of my experiences occurred vicariously while monitoring UHF and HF radios. Through radio communications Crewdogs listened in on what was going on in the rest of the world. Radios enabled them to keep home base informed of progress in completing assigned mission tasks. On occasion they reported inflight emergencies and malfunctions which required the assistance of experts in the command posts. Those so-called experts were supplied with checklists which covered situations Crewdogs might encounter.

From listening to the radios I came to the conclusion that command posts often caused needless stress on Crewdogs. For one thing they were sticklers for details. Their checklists contained silly questions which were religiously asked by the controllers. I have developed my own theory of why that was. This was illustrated on one Chrome Dome mission I flew.

First, let me explain about the High Frequency (HF) radio. One of the duties of the electronic warfare officer was to monitor the HF radio for traffic including messages broadcast for "Sky King" from "Democrat" or other stations in the Strategic Air Command's Giant Talk radio system and to copy the first message received on each mission and any changes afterwards. Democrat was the call sign for the 15AF command post in California. Skyking was the collective call sign for all Single Integrated Operational Plan (SIOP) committed aircraft and missile crews.

We took off from Glasgow and proceeded in an easterly direction on our North Country routing which had us passing south of Chicago and just to the north of New York City to coast-out from the south coast of New England. We onloaded about 110,000 pounds of JP-4 fuel in the Black Goat refueling area which took about 20 minutes of formation flying below the tail of a KC-135A. As a side note here, I noted that some pilots produce great volumes of sweat doing this sort of flying. Other pilots do not. It seemed as though some of them had to work really hard at that skill while for others it seemed like second nature.

Once the onload was completed we proceeded on our flight-planned route, made our left turn just past the Canadian Maritime Provinces and took up a northerly heading. I made the required operations normal radio report after air refueling. This report indicated to the command posts that the air refueling had gone as planned, and we were proceeding on the mission. This would take the flight over international waters between Canada and Greenland. The route would continue all the way to 89° north latitude where another left turn would be made to head for Barter Island on the north coast of Alaska.

After making the operations normal report and then entering grid navigation, things settled down to a normal pace. There was time to listen to the HF radio traffic that was always on. One ongoing HF radio transmission that day originated from a B-52G doing a phone patch to the command post at Seymour-Johnson AFB. I could identify home bases from the call sign lists in effect that day. This crew was flying the Spanish Chrome Dome over the Mediterranean Sea and, shortly prior to my listening in, had completed an air refueling. The crew reported they had noticed unusual fuel panel readings after refueling. As all good crews were expected to do, they called their home command post to make them aware of their problem.

The crew suspected a fuel leak. They sought guidance on what to do next. The person on the B-52G who was doing the talking explained that they had run the checklist and had determined the leak was in the number two pod or nacelle. Their command post listened to all this radio explanation. The very next words out of the controller's mouth were, "Roger, have you run your checklist?" I could sense the anger in the crewmember's response. It was "Roger, we have run the checklist and have isolated the leak to the number two pod." It was polite but stern. After considerable discussion on the details of the fuel leak, the crew requested permission to return to Seymour-Johnson AFB. The request was granted.

I have often wondered about the nature of SAC command post checklists. Were they deliberately composed of stupid and redundant questions just to be recorded on tape recorders? Were these checklists designed to gather detailed information for use by accident investigation boards? I never worked in a command post, but as a Crewdog I suspected their real purpose was to CYA.

Things Happen in the Air And On the Ground

Ken Schmidt

We were scheduled for a night mission with the ultimate intent of night mountainous TA training. Unfortunately, with the passing of time, I don't remember which crew I was on at the time, but I do remember that I was flying in the navigator position and Major Craig Mizner was the IP on board that night. It was to be a typical EWO Profile training sortie with night air refueling, low level navigation and bombing, and a celestial navigation leg on the return back to Carswell. We had just taken off and were headed toward Abilene when the copilot noted that we had low oil pressure on one engine and the pilot team decided to shut the engine down. No sweat, we still had seven engines running. About 30 minutes later, the pilots had to shut down another engine due to another indicated low oil pressure reading. (Still not a problem as this engine was on the other wing—still had six running.) A few minutes later, a third engine had to be shut down, again due to low oil pressure. At that point the mission was aborted and we turned back to Carswell. Since we were too heavy to land, we droned between Carswell and Dyess with gear and flaps down to burn off excess fuel. About two hours into the flight, the upstairs crew noted a "Fuselage Overheat Warning Light" and initiated the appropriate checklists. Since we were closer to home field, we proceeded to Carswell. During the return, the crew suspected a pneumatic leak and the IP came on line and said to "make sure your emergency equipment is in readiness".

The warning in the Dash One stated: "If an air bleed manifold leak is suspected or confirmed, the aircrew should consider all factors involved with the emergency and plan to land as soon as possible at the nearest suitable airbase. A failed pneumatic duct should be regarded as an impending disaster and the pilot should not prolong the flight unnecessarily in the interest of reducing weight or establishing contact with his home base."

Since I was already on 100% oxygen, I tightened up all the parachute and seat belts and unstowed my trigger ring. I looked at the Radar Nav and saw that he had done the same thing. I was a bit concerned, especially about the possibility of making a nylon descent at night; however, we effectively made a safe landing at Carswell.

We later found out that there was actually no serious problems with the aircraft. All the low oil pressure readings and the Fuselage Overheat Warning Light all turned out to be wiring problems. What had the potential to be a serious aircraft accident turned out safely due to the quality training that SAC aircrews received and having a professional pilot team and IP onboard.

BUFF Near the Lake

Again, age starts taking its toll with the memory cells, but I believe this incident happened during the winter of 1979 or 1980. The Fort Worth area was never known for having snow storms that left measurable amounts of snow on the ground. As far as I can recollect during my five years at Carswell, we only had two periods of snow and the worst one is the culprit in this story.

Basically, if there was one snowflake, the area went crazy - people did not know how to drive in the snow and there were multiple accidents. Also there was no snow removal equipment (either for the state and local highways or for Carswell AFB).

Well, we ended up with six inches of snow and all aircraft at the base were grounded until Mother Nature took over and melted the snow. You can imagine the havoc this caused with the flying schedule, especially with a unit that had two bomb squadrons, a tanker squadron, a bomber CCTS and a tanker CFIC school. After two days of no flying, the DO "launched the fleet" to get the flying operations back on line.

My crew got to fly that day and we did not have any trouble with our sortie and getting back to home station. However, one B-52D was not so fortunate. When all the snow that we had started melting, it left pools of water on the runway and at the end of the day, the water stared freezing on the runway. When it came time for the last BUFF to land, the pilot landed a little long on the runway and tried to deploy the drag chute. The drag chute failed to deploy, and the pilot began applying braking application. Due to the long landing, ice on the runway, the brakes not slowing the aircraft fast enough, the aircraft was too far down the runway for the aircrew to apply power and take off again and they were running out of runway. Seeing Lake Worth grow larger in his windows, the pilot nudged the B-52 off to the right of the runway into the grass. The gunner, wanting to make sure he had an exit available, jettisoned his gun turret on the runway.

When the aircraft finally came to a halt, the forward landing gear was in a three-foot high dirt berm and the nose of the aircraft was over the edge of the lake. As I recall from talking to the crew Radar Nav later, after the aircraft had stopped and the aircraft shut down, the navigator opened his hatch and climbed down, only to end up in the lake. I don't remember the outcome of the accident investigation, but quick thinking on the part of the crew kept the aircraft out of the lake.

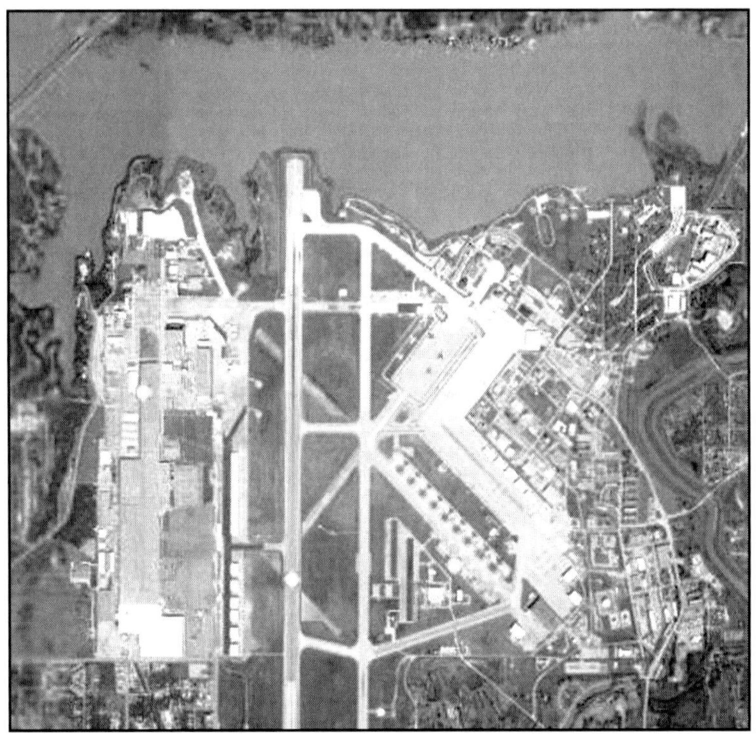

Modern view of Carswell AFB runway showing nearness of lake at the top of photo.

A B-52 Bomber Mid-Winter's Night Dream

Kenneth Boone Sampson

My crew and I had been flying a very grueling and tiring combat B-52 bomber flying schedule at U-Tapao Air Base, Thailand, striking targets in Vietnam, Laos, and Cambodia.

Late in January of 1968 when TET '68 was starting, my crew was assigned a single ship night ferry flight from U-Tapao to Guam. We flew south from U-Tapao, over the Gulf of Thailand, around the southern tip of Vietnam, then east over the South China Sea toward the Philippines. We climbed to 40,000 feet altitude.

The gunner was making no noises on interphone. The electronic warfare officer was silent - asleep in his seat. The radar navigator was asleep in his seat collapsed over the bombsight. The copilot was sacked out asleep on the upper deck. The navigator was monitoring his instruments and navigational heading to the high fix start descent point southwest of Guam. The nav and pilot were lightly chatting about what a beautiful peaceful night it was.

After we crossed the Philippines the pilot invited the nav up to the pilots compartment to sit in the vacant copilot's seat to observe the passing star lit night. The nav at first declined but then acquiesced when the pilot insisted.

The B-52 ASQ-48 bombing navigation system has a long-range navigation function using the pilot's data indicator (PDI) as a heading reference. Before I left my seat I checked the PDI to be sure to point to the high fix southwest of Guam. I planned to navigate from the copilot's seat by watching the PDI centered.

I climbed upstairs passing the sleeping electronic warfare officer and sprawled out copilot. I climbed into the copilot's seat and noted the PDI centered on the high fix. I sat in the copilot seat looking at the stars in the sky, the instrument panel, and lightly chatting with the pilot, with the PDI centered.

Then I fell asleep.

When I woke up, I noted the PDI pegged 90 degrees right while we were in the middle of the Philippine Sea. The pilot's head was slumped on his chest.

I wasn't strapped in the copilot's seat and I didn't know how to fly the airplane. So I reached out my left hand and tapped the pilot's shoulder. He jerked awake and grabbed the yoke to try to pull the B-52 back to altitude, as we were about 4000 feet below our assigned 40000 feet altitude.

As we pulled up, the crew started waking up asking, "What's going on?" I told the pilot to turn right 90 degrees to the PDI and climbed back downstairs to the nav position.

In my ejection seat at my nav position, I plotted our aircraft position and found that we were 60 nautical miles north of course, that we had traveled 94 miles in an arc to the north for a duration of 12 minutes, with everyone asleep, steadily drifting off heading and losing altitude.

Throughout the flight and on the ground nobody ever said anything, and I never flew in the copilot's seat again for the next 300 missions.

My "Dollar Ride" in the B-52

Nick Maier

In 1957 I was the youngest pilot in SAC to enter the B-52 upgrading program. In keeping with Air Force tradition, my crew's first training flight was known as "The Dollar Ride." I was comfortably camped on the jump seat, following the After Takeoff Checklist item for item, and listened intently to the instructor pilot's flying school interphone chatter.

"12,000-foot cabin pressurization check. Reduce your climb Pilot; it appears that the air conditioning pack is malfunctioning. I'll check it in the combat mode. Nothing doing in that switch position either. Lets level off here. Looks like the copilot will have to visually check the pack."

My stomach gave a distinct jump when I realized that the IP was ordering me to enter the forward wheel well and examine the condition of the inoperative equipment. I will never forget the personal decision that I was forced to make after finally managing to open the access door which separated the pressurized crew compartment from the un-pressurized forward wheel well. To wear my cumbersome parachute or leave it behind?

Even with a helmet and headset, the noise level of the area was unbelievable. Besides the rush of air moving over 200 miles an hour around the thin blanket of the aircraft's aluminum skin, I was surrounded by Boeing's infernal air bleed system. The four pneumatic driven alternators, with their screeching, near supersonic spinning turbines and the accompanying pin-hole hot air leaks, added to the high heat environment, making it necessary to be cautious where I placed my hands, even though I was wearing flying gloves. It was easier to move along the cramped crawlway space and get to the pack area, once I took off my chute.

I was forced to lean my full weight on the huge retracted landing gear tires in order to see as much of the inoperative A/C pack as possible. Even then it required a stretch of my neck, with my face surrounded by the smell of hot rubber when it brushed against the treads. There was no visible damage and no apparent leaks. It didn't take me long to exit that Hellhole, get strapped back into my chute, and return to the security of the jump seat. I quickly plugged back onto interphone to give my report.

"The A/C pack looks normal to me, IP. They didn't mention in ground school that the 'Dollar Ride' included an in-flight tour of the forward wheel well." With the loss of cabin pressurization, our crew was forced to abort the high level portion of this first mission, and spent the next six hours in the traffic pattern, practicing bounce and go landings.

After the serious, professional post-mission critique finally turned into a relaxed camaraderie, my aircraft commander casually remarked to his still shaken copilot, "You probably oughta know Ace. While you were wandering around in the wheel well playing crewchief, our illustrious IP suggested lowering the landing gear to cool off the area!"

My mind burst at this dimly heard comment, since my hearing was still slightly impaired as a result of the sum total of the day's deafening clatter. My A/C took a pregnant pause, smiled widely, and then continued matter-of-factly, "I strongly suggested we had better wait until our baby-faced, beardless copilot returned to the cockpit."

Goose Hunting in a B-52E

Arthur Craig Mizner

As a depression area child born in 1935 and raised on a farm in Pennsylvania during World War II with food rationing, family hunting and fishing provided a source of food for the table. The family hunted for small game such as rabbits, squirrels, pheasant, ducks, and geese. In addition, we hunted large game such as deer and bear. My love of hunting and fishing has stayed with me throughout my life. I used small game rifles, shot guns, and large game rifles. My favorite hunting guns are black power muzzle loaders. On 6 October 1966, my goose hunting weapon became a B-52E tail number 57-025. The hunting area was the La Junta low level route in Colorado flying at 270 KIAS (310 MPH) 800 feet above ground level (AGL) at night using terrain avoidance radar.

The B-52E flight of 6 October 1966 was one of 11 pre-planned mission designed by the 96 Bomb Wing Director of Operation (DO) to meet the training requirement of each crewmember position. There was a five-hour, an eight-hour, and an 11-hour package. The five-hour package was a daylight mission for the pilot's requirement. The eight-hour package was a crew package that included the basic pilot activity such as takeoff and landing. In addition, in-flight refueling, one navigation leg (either radar directed or celestial guided) or fighter interceptor activity, high altitude bombing with a Radar Bombing Site (RBS) was scheduled. During the bomb run, the Electronic Warfare Officer (EWO) would defend the aircraft from simulated Surface-to-Air Missiles (SAMs) attacks from the RBS. This mission may also be flown over the water for gunnery training as to firing the 50 caliber machine guns. The 11-hour package included most of the eight-hour package with the addition of a peace time heavy weight take off of 415,000 pounds, one daylight and one night time navigation leg, plus low level navigation using terrain avoidance radar followed by low level bomb runs. Some bomb runs may be flyover, short look, or long look. The five-hour pilot package always followed the 11-hour package. The reason was to get as many sorties out of B-52 before a stand down for maintenance activity. Each crew was required to report the maintenance status of the aircraft before landing. Code One stated the aircraft could be serviced with fuel, oil, and water and be ready within about two hours for the next sortie. The aircraft would be parked in the ready for flight area. Code 2 stated there were some equipment problems with the aircraft but not safety of flight. Details of the equipment problems were called in case the equipment specialist would like to meet the aircraft and look at the problem before engine shut down. The aircraft could be returned to flight status within a short time. The aircraft would be parked maintenance area. Code 3 stated the aircraft had safety of flight problems. The aircraft would be parked in a remote parking area pending repairs and taken out of the flying schedule.

The DYS E-15 mission of 6 October 1966 was an 11-hour package. We arrived at the 337 BS building at 1230 hours to pick up our classified data and personal equipment and then boarded the crew bus for the in-flight kitchen. After picking up our in-flight meals, water, and coffee we arrived at base operation at 1300 hours. To our surprise, a wing staff Lieutenant Colonel IP told us he was getting on board to evaluate to crew. We knew this was a SAC requirement but we were not told before hand. After our weather briefing and filing our flight clearance mission paper work I briefed the Lieutenant Colonel on the mission and his responsibly during an emergency, if any.

After we arrived at the aircraft I reviewed the AFTO form 781 and then briefed the ground crew that this was a staff evaluation of both the flight crew and ground crew. The flight was uneventful with each crew position responding in a professional manner. Before landing we reported the aircraft Code 1. After landing I taxied to aircraft to the ready for flight area and shut down engines. The normal procedure for post flight aircraft inspection on my crew was for me to look at the left side of the aircraft and the copilot look at the right side of the aircraft. As I exited the aircraft last, the Lieutenant Colonel said he only had a few comments for me and rather then going to maintenance and operation debriefing he would talk to in the crew buss. I asked my co pilot to check both sides of the aircraft while the Lieutenant Colonel debriefed me. The condition at that time was very dark and the wings were very high without a full fuel load. In talking with the copilot latter, he stated he only walked under the wings and did not get behind or in front the wing to look at the top of the wing. As this aircraft was scheduled to fly in a few hours, the ground crew serviced the aircraft and called it a through flight that does not require of the inspection of a normal flight.

When the next flight crew arrived at the serviced aircraft for five hour flight package the wings were normal position. During their pre flight inspection the copilot found a very large hole in the wing outboard or the drop tank. There was no replacement aircraft at that time and the sortie was lost.

I arrived home about 5:00 AM and went to bed. Within about one hour the telephone ran and I was told to report to the DO as soon as possible. The DO explained to me we failed to properly perform a post flight inspection and the next sortie was lost. By now, the Director of Maintenance (DM) arrived and he repeated the DO statement and said if the flight crew had found the hole, he could have generated another aircraft for the mission. I then told the DO and DM the ground crew was just as responsible for not seeing the hole as I was. The maintenance crew goes up on the top of the engine to service engine oil and could have looked over the wings. The conversation went down hill from there.

I was then sent to the Safety Office to help fill out the AFR 127-4 in-flight damage (bird strike) report due within 24 hours. The follow up report which included damage, estimated repair hours, and cause of damage was prepared within a few days. The damage was to the leading edge of right wing stations 1267 and 1193-3. Parts replaced: two nose ribs on one leading edge panel. The bird strike was identified by the base veterinarian as a Canadian goose in the 10 to 12 pound range. The Canadian goose remains were wedged inside the wing in front of the main wing spar. The aircraft repaired took over 50 hours to repair. This Canadian goose was not going to be on my dinner table as they were in the past.

We're Going Down

George Donald Jackson

This incident took place over the North Atlantic and I'll only mention one name because he actually saved us. Our copilot was Walt Williams and we were on a 24-hour Cuban Crisis mission that had two BUFFs flying 24-hour missions with H-Bombs and targets.

Our Pilot was a great guy and a good Pilot but he just couldn't air-to-air refuel. He'd always let the staff Pilot take on the fuel if the staff guy wanted to. On this mission we had no staff Pilot but a lot of T-storms and turbulence. The other BUFF got his fuel and stayed with us for a while as we struggled to get ours. The other BUFF Pilot finally said he had to break it off and join the track heading north. We stayed with the tanker until he said he was running low on fuel and had to return to base. Our Pilot thought we had enough fuel to make it to the next tanker over Alaska about 10 hours away.

We turned north and immediately went into a T-Storm. Our radios were filled with loud static that made it impossible to hear anything. I turned all my radios off and had only on my interphone which was clear as a bell. We were all over the sky. The Pilot started yelling. I was the only one who could hear the interphone because everyone else still had their radios on. I could hear that the Pilot and copilot were fighting over the control of the throttles. The pilot's air-speed indicator was winding down and he was trying to increase the power by pushing the throttles forward. The copilot was shoving his hand away and pulling the throttles back.

"We're going down!" the Pilot was yelling.

That was not something you wanted to hear coming over the intercom when you are in an aircraft in the middle of a thunderstorm. That was especially true when the flight was one of your first ones after getting back onto flying status after a lengthy hospital stay as a result of having to eject from another aircraft that was going down.

Again I was the only one of the rest of the crew hearing this due to the radio noise over everyone else's headsets. The copilot was pointing to the instrument panel and doing his best to take control of the aircraft. The Pilot finally said, "OK, copilot you have it - fly fuel flow." He had thought the copilot had lost his air speed indicator also. The copilot took control of the airplane and got us through the T-storm. The static on the radio finally subsided and everyone could then hear.

It was one of my first flights back after being grounded for a year and the Pilot knew I was a bit nervous! HA! The Pilot then said in a calm voice, "Don't worry EW. We'll get you through this."

I replied, "We're over the North Atlantic - where do you think I was going to go?"

Anyway we made it through that one and a few more incidents before the 99th Bomb Wing broke up and I went to Walker AFB. Then it was off to Westover and a few trips to Nam, stories which I covered in "*We Were Crewdogs II – More B-52 Crewdog Tales*."

What the Hell Was That?

Harold E. Pfeifer

Some of our B-52E aircraft had to be returned to the factory at Boeing Field in Seattle, Washington for minor modifications. Our crew (A/C Richard H. Avery, CP William K Hilton, RN Harold E. Pfeifer, Nav Frank J Emma, EWO Jerry B. Hancock, and Gunner, Wilbert C. Moore) departed Altus AFB, Oklahoma in early September 1958 to take a B-52 to Seattle. Avery had been a B-24 copilot who survived the initial Polesti oil field raids in WWII.

We had been flying and climbing about 15 minutes. All of a sudden the B-52 started a strong roll of about 45 degrees to the left. A pilot announced over the interphone "What the hell was that?" The A/C immediately stopped the roll, and rolled the aircraft level from the climb. The A/C stated the airplane seemed to respond to the controls, and added that he didn't think we had a midair collision. After a few "quick" minutes, three crew members reported no more usual happenings; however, the tail gunner reported something had struck the vertical tail above him, and seemed to be gone. Our crew talked the situation over, and since the B-52 was flying OK, decided to continue to Seattle.

The CP called Altus Tower and explained that something had struck the vertical tail of the B-52, a strong roll to the left had started, but control was regained immediately, and the airplane continued in level flight for a few minutes. The A/C reported we would continue on the mission.

During the following hour of flight, the EWO installed the sextant and scanned the top of the airplane for damage. He reported the ejection seat lifter cover appeared to be missing. The cover is about two feet wide, four feet long and about six inches deep. More study of the top rear of the B-52 showed no other apparent damage. The CP relayed to Altus that the RN estimated that the lifter might be found about 20 miles on a NW heading from Altus AFB. In fact, the cover was picked up two days later.

The rest of the flight was routine. After we landed and taxied to a parking location, we were greeted by many vehicles and people walking toward the airplane. We exited the airplane, walked to the tail area, and looked at the damage. The vertical fin had a horizontal cut that was six to eight inches high and located about five feet above the top of the B-52 that was two feet into the fin. We heard later the damage was minor.

Shortly after, our transport aircraft arrived and we returned to Altus AFB, Oklahoma, leaving the B-52 at Boeing for repair.

Inflight Emergencies

Jimmy Turk

My first B-52 inflight emergency occurred on 21 August 1961 in B-52F 57-037 during our CCTS flight #7. The crew consisted of pilot Capt Roger L. Harris, copilot Capt Richard (NMI) Powell, RN Capt John (Jack) L. LaHaye, EW 1 Lt James R. Wiley, and gunner MSgt Marvin L. Sipes, with IP Maj Charles Smith and IN Major Chilton, and myself a 2nd Lt as Nav. We were very fortunate that our instructor crew each had over 8,000 of B-52 time and had been the core of one of the three B-52 crews that completed the non-stop around-the-world flight.

It was a 0325Z (9:25PM) standard night take-off to the north on a Chowchilla #1 departure. From lift-off the aircraft demonstrated horizontal control problems with periodic shifts to the left or right. Passing the 12,000 foot oxygen check, the IP ask both my pilots if the shifting was detectable in both yolks. They both said yes. The IP asked the copilot to contact departure control for clearance to a holding pattern west of the base where we maintained 24,000 MSL and 250 IAS as we all tried to determine what was causing the random shifting that was very pronounced in the gunner's position.

Major Smith in his over 8,000 hours B-52 time had never experienced this type of phenomena. The IP contacted the command post for a phone patch to maintenance. The one hour plus exchanges between the on-board crew and the maintenance and command post personnel had us checking every conceivable combination of control surface checks, hydraulic pack recycles, multiple circuit breakers pulled and re-set – all to no avail. The command post finally directed us to stay in the local orbit and burn off fuel until we were under 250,000 gross weight and then to land. The IP switched seats with the copilot and we proceeded to lower the gear and run the air speed at just under red-line for that configuration to generate a fuel burn rate of about 40,000 pounds an hour.

The IP declared an in-flight emergency prior to starting our departure out of the holding pattern. That provide the entire Castle AFB Crash and Rescue team to respond and respond they did. There were fire trucks, crash and recovery vehicles, ambulances, security police, and the base operations Supervisor of Flying (SOF) officer all along some part of the runway and taxiways. The IP in the copilot's seat made the landing seem uneventful. As Captain Harris said at de-briefing "It looked like a huge Christmas Tree as we approached and landed - thank God we didn't need them." The IP and the crew met the next day and in-mass checked with maintenance for the cause of the emergency. The problem turned out to be a loss of a magnetic sensor in the rudder, causing it swing back and forth at will. Another set of notes and cautions were added to the maintenance procedures.

The same basic crew on our seventh B-52H and fifth solo flight after we arrived at our new home station of Kincheloe AFB, Michigan experienced a Dash-1 Emergency Tech Order change when the following occurred. The flight actually originated out of Wurtsmith AFB, Michigan when we were bringing a new plane home that had been reallocated from the 379theBomb Wing to us. The 15 December 1961 afternoon take-off was at 1730Z (13:30 EST) in 60-053 for a regular crew training flight. It was a normal training mission profile takeoff, A/R, High Level celestial navigation leg and a return to our home base of Kincheloe, During the departure and prior to refueling we lost the #5 engine and the associated generator and hydraulic pack. With the B-52H having 120KVA constant speed drive generators - three were quite sufficient.

The engine continued to windmill and that provided operational hydraulic pressure of all air break segments. The pilot remembered that on older B-52s the loss of hydraulics on one side required us to take the sister system on the opposite wing off line. Our Dash-1 had no such warning. So just to be cautious, we made a practice approach at 20,000 feet over Kincheloe and all was fine with full air brake and lateral control. We did another check at 5,000 feet with the same result. A final pass was made at 1,000 feet above the runway and all was well. Everything remained normal until we were about 250' AGL and the air speed dropped below 140 KIAS. The number five engine stopped rotating, its hydraulic pressure went to zero and the lift dynamic was destroyed as the aircraft started to roll right. Both pilots fought to upright the aircraft and did so just before touch-down. Yes, part of debriefing was the documentation of a Safety-of-Flight procedure to add the pulling of both pairs of hydraulic CBs to maintain safe flight control. The procedure was distributed to all B-52H Dash-1's within 48 hours.

Chapter Four

Southeast Asia [south-eest] [ey-zhuh] - *noun* - The countries and land area of Brunei, Burma, Cambodia, Indonesia, Laos, Malaysia, the Philippines, Singapore, Thailand, and Vietnam.

The author in Thailand.

Memories of a Former Crew Dog – Bored and Lonely

Doug Cooper

The things that I remember about my Arc Light/Bullet Shot experiences are being lonely and being bored nearly to death. Lonely for home and family (I'd been married less than six months when I was tapped at Beale for RTU) and bored with an existence that consisted mainly of the same old 12-hour flights day after day. I think the lonely part broke up a lot of marriages. Mine survived, to this day, mostly due to my wife's patience. It prepared her well for unaccompanied tours later in our career.

The flying was really boring. Get up, shower and shave, eat something and go to briefing. Pre-flight, wait around to start engines, taxi, and takeoff. Bore holes in the sky for about five hours before we hit the tanker northeast of the Philippines. Then cruise into South Vietnam and drop the bombs at the direction of some guy on the ground on some suspected enemy supply depot or camp. Turn east and head back to Guam, another six hours away unless you had "hangers" (bombs not dropped) and then you had to avoid the Philippines which added a couple more hours. The job had all the excitement of being a long-haul truck driver without being able to stop for coffee. Speaking of coffee, the in-flight kitchen coffee had to be the worst every brewed. But it was the only coffee available so we drank it anyway. The facilities were sparse and you didn't ever want to be the first one to use the "honey bucket" because you then had to clean it out.

Gilligan's Island – a favorite!

Finally, back on the ground, all you wanted to do was sleep after the obligatory chili dog and beer at Gilligan's Island; however, if you succumbed, you had to sleep 16 hours because that was when you had to meet the next briefing. Continue doing this until you ran out of flying time. Then you got to be duty crew or pre-flight crew, both really invigorating activities. If you didn't have the mandatory Seiko watch, you forgot what day it was.

Recreational options on Guam were limited. The acronym (G.U.A.M.) stood for "Give Up And Masturbate" or something like that. There was the BX or the Navy BX, the movies, Kenny's Steak House, the O' Club, Taragi Beach, or just lying on your bed reading. I think everybody read out of self defense. Almost everyone drank, also out of self defense.

Kadena and U-Tapao were more interesting, but after a few days there, the situation was much the same as Guam especially after repeated trips to the forward bases. At least the flights were shorter. And, at Kadena, you could at least get a meal anytime and the BOQ's were nice. At U-Tapao, we lived in trailers until the dorms were completed. The last time I was in Thailand, I didn't leave the base.

The boredom continued until the Christmas party we all had in 1972 when things got really exciting and the crew dogs accomplished what should've been done in 1965. If you weren't a cynic by then you quickly became one when the politicos from both sides couldn't decide on the shape of the peace tables. The end of the war party was kind of like a premature ejaculation.

I finished up with five tours. After Beale, I went to Carswell just in time for Bullet Shot. Number five was substituting on a Kincheloe crew for three months while we flew training missions out of U-Tapao. If I never see Guam again, it will be too soon.

I still admire the tenacity of the Crewdog. It's amazing that many of us survived those years with a reasonable degree of sanity. We all left something over there. Some of us left our youth. But some left their lives. I think about them often.

Memories of Senior Staff
The War is Back Here Now

Finally, we got all the airplanes and Crewdogs back to Carswell. SAC couldn't wait to reconstitute the Alert Force. After a couple of years with only a few H models pulling alert, suddenly we needed to get the nukes back on all the G's and D's ready to signal the end of the world.

The world situation must be worsening. Were the Soviets rattling their swords? Were the Chinese getting ready to move into Taiwan?

No. We needed to get back to being combat ready. We needed to have an I.G. visit.

Major Tom O'Malley, our ops officer, put in very succinctly. (I'm paraphrasing.) "Guys, the real war is now here in the States. Get used to it."

We Need to Launch a Tanker

The time is 1968. I am a 1st Lieutenant, half in the bag, sitting on the bench in front of the Andersen AFB O' Club. I am waiting for the Red or Blue Line Bus (I forget which) to take me back to the compound. My bench mates are two old Lieutenant Colonels.

Lieutenant Colonels ran the operation. The full bulls were just there for the brigadier general qualification points and an Air Force Cross or two.

Lt Col #1, "If we have a BUFF come back scosh on fuel from a combat mission, how are we going to get a tanker up in time to meet them before they run out of gas? Do we have any procedures?"

Lt Col #2, "Damned if I know."

Fast forward about five years.

I am a captain now. Same bench. Different Lt Cols.

Lt Col #1, "If we have a BUFF come back scosh on fuel from a training flight, how are we going to get a tanker up in time to meet them before they run out of gas? Do we have any procedures?"

Lt Col #2, "Damned if I know."

The more things change, the more they stay the same.

Prisoners on Guam

Many Crewdogs will remember the efforts of some of the folks to get our POWs back. There were the POW bracelets, POW bumper stickers, POW book marks.

After the war was over, there were still a lot of Crewdogs on the Rock just in case we had to go back and drop some more bombs. An idle Crewdog is a force to be reckoned with.

Somebody got the bright idea to show how Crewdogs appreciate being confined to the Rock by emulating the POW bumper stickers with a "Free the Prisoners on Guam" sticker. I know who originated the POG idea but will not rat this person out for less than $1.95 in small bills.

Anyway, the stickers (they were printed on sticky-back paper) appeared all over the island, on buses, staff vehicles, private cars, latrine stalls and other places too numerous to describe.

I think they assigned a Lieutenant Colonel to find out who the culprits were. They also issued him a scraper and he could be seen about the base busily scraping the evidence into garbage bags.

He could've waited. The stickers succumbed quickly to the continuing tropical rains and the ultraviolet tropical sun rays.

Extended Alert

After getting back from my first Arc Light and getting recertified on the EWO, our crew got back into the endless cycle of alert duty. At Beale, we had three and four day alerts because the base was considered "remote." The alert shack was at the end of the base, miles from the BX and the Club so we could only go as far as the squadron building while on alert. The family visitation center was an outdoor patio so visits were only possible during fair weather.

Sometime during the late 1960s, all of the BUFFs went into an extended alert. All of the aircraft were generated and crews recalled. Crew dogs were billeted wherever there was room. The alert shack was like a Salvation Army shelter. As with any group of Crewdogs left to their own devices, things degenerated quickly into a huge game of grab-ass. No one was safe. The great effort went away after a month or so. Many of us went non-current and had to be checked by instructors who also were non-current.

We never found out why or what crisis precipitated the build up. Rumor was that it was due to the B-58s being retired or trouble with the Titan Missiles. Probably never know.

Escape of the Crew Dog

In 1974, I got back from one of those all-nighters and stopped by the squadron on my way home. There was a handwritten note on the door of the squadron commander asking for anyone interested in going to the Command Post to see him. I thought, "What the Hell?" and waited a couple of hours to see Lt Colonel Cornelius. He said he'd put my name on the list and forward it to the DOC. I thought, "so much for that". I wasn't a pilot and that seemed to be an impregnable obstacle.

A couple of days later, Lt Col Johnny Rogers, the DOC, called me and I went in to what turned out to be an interview. I got the job. I couldn't believe it.

Apparently no one else could. It took two years to get an approved job description, called an AFSC – I think, in the personnel system. I was carried on the rolls as an instructor in the CCTS even though I worked a full schedule in the C.P. On my first shift as a command post officer controller, the SAC Senior Controller briefed the CINCSAC that he had a non-pilot officer controller in the Carswell Command Post. This relayed to me by the senior staff some days later. He said the CINC only shrugged his shoulders.

My eternal gratitude goes out to Lt Colonel Cornelius for having enough faith in me to forward my name. But my eventual success credit goes to Johnny Rogers, a great officer and a true gentleman. I will always think of him as my mentor.

I never looked back. I didn't miss the BUFF but I have a lot of respect for those Crewdogs who flew the "heavies." It was never an easy job and never had the glory that the fighter jocks had.

Guam Moments

J. J. Parker

It was during the Great Cambodian Air War, in April of 1973, and we were flying a nighttime three-ship cell out of Guam. I was "New Guy" copilot when I asked the A/C, Dave Chicci, about the sleeping arrangements - such as who slept and when. Well, before I tell you the rest of the story, I'd better go back to the middle, or the end of the beginning.

I chose to fly a B-52. I wanted to go kill the "Godless Commies", so I got my chance. Like many others, I flew to Guam on a tanker out of March AFB, California with an engine or two suspended and strapped in the middle of the aircraft's cargo compartment. I was in the company of guys trying to make their nests for the long journey all around the thing. It would be an understatement if I said it was crowded back there on the way over. For me, that was not a bad thing. I was pretty stupid back then, and might still be, but that's for another book. Anyway, I figured, "Here I am in warm California, and I'm going to also warm Guam." I get onto the plane with my B-4 bag, maybe a helmet bag, and wearing my 1505s - the short sleeve khaki shirt, etc. uniform. You're right, no jacket; no nothing that was warm. My bag joined the other cargo that was all stacked up and tied down somewhere in the rear of the KC-135. Soon we're at 31,000 feet and I'm getting more than just a little cold. Darn, that was a miserable flight! I ended up having to huddle close to another enroute copilot whom I came to know quite well during the darn long flight.

Finally I get my feet on ground in Guam and there's a fair, humid breeze. "Ahhh, what a nice place. Warm!" I meet my new crew who have all been there together through the worst of it. I'm replacing their copilot who is going back to the states to upgrade. I'm the new guy. Right away, I notice one thing about the whole bunch of them. Their flight suit name tags are not like any I had seen before. Their name tags had some strange letters and numbers right below their names which looked like a "D", an "O" and an "S". Yup, that's what it was, DOS. They all had their dates of separation from the USAF, right there up front, just in case anyone cared. Imagine my excitement when I, a

1st Lieutenant with about three-plus years before my commitment was up, was handed my new "crew" name tag. Right below my name, there it was: DOS 10 JULY 1977. I was so proud. I actually felt I'd been accepted.

Try to imagine how much a BUFF crew had gone through together over there during Linebacker II, or the Haiphong raids, and how tight they were because of those events. Then I show up, wanting to go fly even before SAC Contingency Aircrew Training (better known as SCAT) school started for me (which I did, on some guy-from-somewhere's D-model). Yup, "Mr. Gung-ho, meet the DOS gang!" But soon enough all was well, or as well as a Lieutenant could ever know. We were soon going to fly our missions, and be taxi crew, and spare crew, and play chess, and all that stuff.

I'll get back to the sleep arrangements on that first mission with my guys in a minute, but first, how about those Combat Mission Folders and target study and all that other pre-mission stuff? Guess what? There were lots of brass there where that was all going on. Needless to say, one colonel asked me if I thought I was a smart *ss, and I had a hard time answering him, because, of course, I was. But how'd he know? I hadn't said anything to him. Oh yeah, the DOS name tag! Clearly, he didn't see the humor in it. I wore it that whole tour, mostly because I had no appreciation for what message it was sending to anyone more senior in rank. For the most part, if my memory serves me, most Majors and below laughed, Lieutenant Colonels looked sort of sideways at me, and Colonels just glared. It's easy being stupid when you don't know any better. Darn, I miss those days!

Okay, so we're heading out to the G-model for my first flight with the guys. Of course, it was a night three-ship cell mission and the idea of crew rest was pretty much something for older folks. Know what I mean, Vern? But, I was feeling a tad tired, and so I asked Dave about our eyelid rest plan. He said he'd sleep first, and then he took out his grease pencil (wonder if that grease pencil company went out of business when those Vis-à-Vis pens came along?) and marked a spot over each blip on the old Terrain Avoidance (TA) scope. Now, the old TA scope was round, about 5"-6" in diameter, and it had a Plan view, and a Profile view. In the Plan mode, it had a display like a typical radar screen of those days. It was like you were looking down at the world, and the sweep would go by, left to right and back. You'd see blips, or tops of terrain, etc., depending upon how you had it all set up.

At any rate, he told me to keep the other two blips (BUFFs) right in the same relative location as the marks on the screen. I was diligent and did so. I was working ever so hard to keep those things right where he'd told me, just in case maybe he was sleeping with one eye open and was checking up on me or something. He woke up a while later, and told me to tilt it back and take a nap, which I proceeded to do.

I was awakened quite rudely. All I could hear was yelling and all kinds of noise on the radio. As I came to my senses, I heard our call sign frequently being used. People were calling us and when I looked at the TA scope, there were no blips anywhere. Then I looked over at Dave, and he was sound asleep, but he wasn't alone - everyone in our plane was asleep and we were now lead ship and somewhere way out in front of the formation!

Welcome to the real world "New Guy"! These situations aren't covered in any SAC training manual.

I didn't sleep much on missions after that.

I still have that DOS name tag, by the way.

Arc Light Tour from Hell

John York

The year was 1972 and I was a brand new B-52G Aircraft Commander. I could climb any mountain, slay any dragon, and defeat any enemy. After all I had finished pilot training, flown the O-E Birddog in-country for a year and spent my obligatory stint as a copilot on the BUFF. Now following a quick in-unit upgrade I was looking forward to having my own crew with all the trials and tribulations that go with it. I expected to pull alert, fly an annual ORI and maybe even face the devil when the evil CEG made its annual calling. Little did I know what was to come or how rapidly this cocky young pilot would come to his knees.

Following my initial Aircraft Commander Standboard check and certifying the unit EWO and ORI missions I reported for mission planning for the first flight with my crew. I knew I had been blessed with a great crew: CP Jimmy Grisson, RN Steve Shick, N Jim Garrett, EW Bill Watkins, and Gunner Vince Valpariso. They were all well trained and highly competent. What more could I ask for? As soon as I drove onto the base that springtime morning I knew there was trouble. The entire base was going crazy. People were moving in every direction. "Must be the ORI," I thought. Well, no problem, I'm ready for it. We'll take the tests, fly the mission and score shacks on all the bomb runs. This can't be any big deal at all.

"Not so fast," the friendly Operations Officer advised me as I entered my squadron briefing room. He quickly placed TDY orders in my hand while mumbling that the entire Wing was being deployed to Guam and that I should quickly round up my crew and get ready to deploy along with a few thousand other people. "Now this is really action," I thought. "Never been to Guam, it can't be as bad as those experienced guys make it out to be. This is going to be fun." Oh, the innocence of the young.

The Young Tiger KC-135 we were scheduled to catch to Guam didn't materialize for a few days. When we left we spent an unexpected night at Beale AFB followed by another unscheduled RON in beautiful Hawaii. I had been cautioned by the experienced crews to expect a rather vigorous haircut inspection upon arrival in Hawaii. It seemed that a colonel had taken it upon himself to see that EVERY crewmember transiting the island strictly confirmed to AFR 35-10 and especially the haircut standards. He didn't spend much time with us since we all had fresh haircuts. The story was that he would be much tougher when we encountered him at the end of our Arc Light tour.

Our arrival on Guam was noteworthy. Because a number of B-52G wings had been deployed at the same time, the entire operation was disorganized. The D-model guys were there and went about their everyday business but it took a week or so to get the rest of us organized for our first flight. The two required instructor flights went well and soon we were on our own. Now I don't claim to be a smart person, but I knew enough to know that I should just work hard and keep a low profile. There were so many different units there that SAC had a hard time creating the infrastructure. Finally it was decided that all the TDY crews would be assigned to one of the various squadrons which came under a Provisional Bomb Wing. Somehow I managed to be assigned to a unit commanded by a person I shall call Colonel Dastardly. He had a perpetual scowl on his face and if he ever said a kind word to anyone I never knew of it. Little did I know what was in store for me. After all, there is the age-old cliché "To err is human; to forgive is not SAC policy." Because all the D and G-model wings were there we had more colonels than should have ever been allowed. The joke at the Officers' Club was that there was a different full colonel in charge of the launch of each cell that launched out of Anderson AFB.

Things went smoothly for the first couple of months. Our fine crew quickly moved to cell lead position and the RN and I were put on instructor orders. We flew "over the shoulder" flights with newly arriving crews. As the age-old saying goes: "Up until the IP everything looked fine."

Early summer brought a short break in the flying schedule and our crew enjoyed a few days of R&R in Okinawa and Bangkok. It was during this break that I learned of the Anderson G-model accident where the crew bailed out into a typhoon and resulted in the death of my good friend Arkey Vaughn. Arkey was a fine man and we had enjoyed many post-mission beers together at Gilligan's Island and the Anderson O' Club. Arkie was older than me, but he shared a lot of experience and I depended on his wisdom.

We quickly settled back into the pace of the Arc Light schedule and I felt very confident. Then during a flight as we neared the coast of Vietnam the Gunner opened the Personal Equipment (PE) Box to get out our equipment only to discover that we were missing a survival vest and a handgun. I couldn't believe what he was telling me and went back to his station and counted for myself. The Gunner sat in the main cockpit in the G-model. Yep, we were one short. I ordered the rest of the crew to don a vest and I went over the target without one. I knew, however, that when we got back to Anderson there was going to be hell to pay. You don't lose a weapon in Uncle Sam's Air Force and get by with it. It just doesn't happen.

Upon arrival at Anderson we taxied in, shut down our engines, and took the crew bus to mission debriefing before going to the Life Support building where we turned in our PE box. In theory our survival vests and handguns were securely locked in this box because it had the crew's padlock on it, so no one else could enter it. This would be the subject of much debate in the coming days. Finally I made the call to the Security Police that we had "misplaced" a handgun. As you might guess that brought a pretty quick response and resulted in several hours of what I'll politely call "interviews."

The following day brought more interviews with various levels of interest to include a couple of the colonels who had no more responsibility than to see that their "assigned cell" launched on time. I had at least two visits from OSI agents. I never was successful in explaining to them how the "tail gunner" (their words not mine) could reach the cockpit during flight. Several people asked me if we (the crew) could have simply left a vest and pistol in an airplane on the previous mission. I knew that was a possibility but preferred to look for an easier answer. I realized that I needed some reinforcement here and I called my home base squadron commander who was also deployed and commanding a different squadron than mine.

Lt Col John Davis was a great person and he saved my bacon more times than one. He came to see me and told me that he had already heard the story and that he would try to help. He also advised me that Colonel Dastardly was considering punishing me under Article 15 of the UCMJ. After all the A/C is responsible for everything that happens involving the crew. I decided that we needed to look at the PE box so off to Life Support we went to examine it. With a stroke of brilliance Lt Col Davis took a pair of fingernail clippers out of his pocket and pushed the pin out of the hinge and, low and behold, we had just proved that the box was not secure during the time it was left at Life Support. A call to OSI brought the agent who was still scratching his head and trying to figure out how the tail gunner got into the cockpit. We again demonstrated to him how the PE Box was not secure even with the aircrew's padlock on it. Lt Col Davis saved me and I will always be grateful. As to what really happened to the survival vest and handgun – well, I stand by my story.

I knew that I had dodged a bullet and I talked sincerely to my fellow crewmembers about the fact that we needed to stay out of trouble the rest of our Arc Light tour. Things went well for a few weeks as the war started to heat up a little. One night we were assigned to bomb a target well north of the DMZ near the city of Vinh. We were scheduled to have fighter support and an EB-66 just off the coast providing ECM support. The standard procedures required the EB-66 to broadcast any SAM launch and to get an acknowledgement from the bomber crews. Once again "Up until the IP everything looked fine" and as we flew the mission we rolled out for a synchronous bomb run. The bomb doors were opened at 60 seconds to go when the Nav reported "There's something strange on the scope. I think it's coming up at us." Shortly afterward I heard "Three has uplink" followed immediately by "Two has uplink" Then our EW stated over the interphone "Damn it Pilot, I thought so. They've launched a SAM at us."

Now our instructions for this mission required an immediate divert in the event of a SAM launch and that's exactly what we did. The only problem was that I had forgotten the code word for "divert" so I made the only transmission I could think of: "Red Cell, let's get the f*** out of here". Everyone understood what I was saying. The SAM followed our aircraft through the turn and passed slightly off the nose. Fortunately the proximity fuse didn't work properly and it exploded well above our altitude. We proceeded to bomb a secondary target below the DMZ.

Three days later I was standing in a brace in Colonel Dastardly's office and he was not a happy camper. "How could you embarrass SAC like that?" he asked. I decided to keep quiet and see where this was going because I had no idea what he was talking about. He then went on to say that Seventh Air Force had filed a complaint because the EB-66 accused the bomber flight of leaving them alone over the target. Now I knew that the orbit which the EB-66 had been assigned was well off the coast of North Vietnam, besides the fact that he was there to protect us, not vice versa. I also knew that Colonel Dastardly was in a particularly foul mood so I said "Yes Sir" to his every comment or question. I was summarily dismissed and feeling pretty down in the dumps. How could I ever hope to have a career in this man's Air Force the way things were going? Oh well, I was still young and only had a couple of years left on my obligation.

I left there and went to Pen Aids where I pulled the audio tape of that mission. What I heard on the tape knocked my socks off. The recording revealed that about the time of the SAM launch the EB-66 made the following transmission "Jimmy 24 has Uplink." Obviously no one in the bomber force heard the transmission because it was not acknowledged. Neither was it repeated by the EB-66. So the true story was that the EB-66 was safely off the coast out of the target area, knew of the SAM launch but did not ensure that the bomber force was aware of it. And they had the audacity to accuse me of going off and leaving them unprotected. Colonel Dastardly would never know of this though; I'd already had my butt chewing.

Finally we neared the end of our Arc Light tour. The end was in sight. I was hoping that I could complete this tour without totally ruining any chances of an Air Force career. All I had to do was fly a half-dozen more missions then I could go home for 28 days before returning for another tour. By then Colonel Dastardly would be gone and all my troubles would be over. I showed up early for every flight, made sure I had good target study, and really toed the line. I even stayed away from the Officers' Club at night. A few days before our flight home we took care of the administrative things involved. Some of the guys had bought glass elephants and habachi pots (big bar-b-que pits) that needed to be crated and shipped. We got fresh haircuts in order to get by the "Haircut Colonel" during the scheduled refueling stop in Hawaii. A Chief Master Sergeant from my home base came by and asked me to escort a box of electronic components for him. He said that they weren't needed at Anderson AFB and needed to be returned to home base. He stated that he would take care of having the box loaded on the airplane and picked up at home base. That seemed like a responsible request and I agreed to do so.

The big day finally arrived. We had flown all our missions and the only thing that appeared to be damaged was my career. How much so was yet to be determined and I was modestly hopeful. We gathered at the passenger terminal about two hours before scheduled departure time. As we neared departure time I was paged over the PA system. Low and behold, it was the OSI agent, the one who was still trying to figure out how the tail gunner got into the cockpit. He advised me that my luggage had been dropped while being loaded onto the airplane and that liquor came pouring out. He wanted to know why I was trying to smuggle liquor back into the US. Now I knew that I didn't have any liquor in my luggage and asked to see the offending baggage. We walked out to the airplane and there was a wooden crate marked "Electronic Components - Handle With Care." It had my name written on it. That Chief Master Sergeant set me up I thought. As luck would have it a staff car drove by and the driver saw all the commotion and came up to see what was going on. To my horror it was Colonel Dastardly. He asked what had happened and I explained it in the simplest way I could. I'll never know what caused his out of character response because he seemed to take great delight in my story. He smiled, told me to go on and enjoy my time at home; that I had been through enough.

I was grateful for his encouragement and indeed I tried to enjoy every minute of my short time at home before returning to yet (gasp) another Arc Light tour.

Charlie Tower

George Golding

After flying two Bullet Shot tours including Linebacker I & II, I was back at McCoy AFB, Florida wondering where we'd be sent next. We had heard that the base was closing and everyone was just waiting to get orders. I got a call to report in and I thought here it is. Well it was orders, but to go back to Andersen Air Force Base, Guam again only this time in a staff position. I was going to be a scheduler, but when I got there I ran into some old friends who were looking for a "D" Charlie. A Charlie had to have 2,000 hours in the B-52, be an Instructor Pilot in the "D" model, and be able to handle up to 15 or more aircraft simultaneously while talking to the Control Tower, the Wing Commander, the Command Post, HQ SAC and the Boeing experts in Wichita, Kansas all at the same time. You also had to be a field grade officer with at least one tour flying missions out of Andersen, and be lead aircraft qualified. The next thing I knew I was hired and placed on 12-hour shifts, six days a week. After three weeks in the tower you got a three-day break which would let you fly a mission in order to qualify for combat pay and then change from day to night shifts or vice versa.

It was Charlie Tower that talked to the aircrews from the time they arrived at the aircraft until after they were airborne. In the tower, located between the two active runways, all communications were handled by the two Charlie's on duty. One was a "D" and one was a "G", representing the two types of B-52 models flying combat missions out of Anderson AFB. They also handled the KC-135's passing through.

Communications consisted of UFH radios and FM radios along with dedicated phone lines to the Command Post, Tower, the Fire/Crash Department and Ops. Charlie Tower was manned by the two flying Charlie's, a Super Charlie, who was a 0-6 Deputy Ops Group Commander, as well as Uncle Ned and Cousin Fred. Uncle Ned was a Maintenance Officer who helped with aircraft problems and Cousin Fred was a transportation NCO who handled all the buses and ground transportation for the flight crews and all the maintenance folks. The senior Charlie wasn't always there as he had other duties but, if a big operation was going on, he was there. We knew something was up the more crowed the tower became. I have seen it with standing room only with everyone from the General on down. Still, it was Charlie, who talked to the crews. After all, we had been in their shoes and in most cases had encountered the same problems. Everything from bag drags, moving from a broken aircraft to a different one, to replacing a black box or even a crewmember with the aircraft on the hammerhead within minutes of scheduled take-off time. Whatever went wrong, the answer to the problem was handled by Charlie. The maintenance folks had a van, called a "Red Ball", which consisted of technicians and parts all ready to rush to an aircraft that was having a maintenance problem. These were the most qualified and highly dedicated NCOs in their career field. The most important requirement was, as always in SAC, an on time take-off.

We kept score by the number of "ball games" launched with a full cell of aircraft. A cell consisted of three aircraft and each cell had a different color for its call sign. It could be Red Cell or Green Cell or Cobalt etc. Red One would be the lead airplane in the cell, followed by Red Two and Red Three. Due to maintenance problems, and there were hundreds of them, or another cause for delay, the crew in Red Three could become Red One if they took off first. Charlie's job was to get the Ball Game off on time. At times there was a new Ball Game every 15 minutes. With a 45-minute before take-off engine start time and a 30-minute before take-off taxi time Charlie could have up to 15 or more aircraft taxiing counting ballgames that had landed at any one time. Trying to keep the entire field straight was a very busy job.

Charlie's biggest fear was to have two aircraft taxiing head to head with no place to go. If an aircraft in a cell taxiing out for take-off turned the wrong direction out of his parking spot it could delay their ball game and, at times, another one as well. One rainy night at 0-dark-30, a new crew's first mission in Silver Three, and a brand new copilot did just that. Charlie Tower told him to turn right out of his spot and the young copilot got his hands mixed up and turned left, right into the path of Silver Two. It took over 25 minutes to get a tug over to them and backed into their parking stub so they could let Silver Two get by and then get out to the runway. They did get off on time, but the copilot was forever called "Lefty" and Charlie had heartburn the rest of the night.

Handling airborne emergencies took first priority. Whether it was an engine failure after take-off or a problem prior to landing, Charlie was there to help the crew. I remember a double engine failure right after take-off. We had the crew jettison the weapons in the water and put the aircraft in a holding pattern until they could burn off enough fuel to get light enough to land. Since the aircraft's weight was around 450,000 pounds at take-off it had to be down to about 225,000 pounds to land, that's a lot of fuel to burn off since the B-52 did not have a fuel dump capability. On another occasion a returning B-52 could not get its left forward main gear down and we talked via phone to the experts at Boeing in Wichita, Kansas to help with the problem and the gear was finally lowered successfully. There was a time we helped a crew right after bomb release when they could not get the bomb doors closed. They called the Command Post via HF radio and we were connected to the Command Post via a dedicated phone line. For any problem from before engine start till engine shutdown Charlie was there to help

We were able to relieve some stress and also had fun in the tower. We graded landings when it was quiet and all the full bulls (colonels) were asleep or off the air. We had a system that was included in the "Falcon Code" list. That list was an unauthorized document, but almost every crew had one. For example you could give a crew a Falcon 124, which was as I remember, "Take that runway, and that and that!" Needless to say it wasn't a smooth landing. There were over 30 codes used throughout Southeast Asia. Some were only used at certain bases and by certain crews. Charlie also had music to play on certain occasions, like "Walking in the Rain", or to announce the "crack" of dawn. Needless to say it was never dull. It was an honor to work there and with the finest aircrews in the Air Force.

Charlie on the Job

It blew extremely windy on the Isle of Guam that day,
With four degrees of crosswind crab affecting all the play.
And so, the crews came to the aircraft, their faces wet with sweat,
And sat completing checklists, with flight suits wringing wet.
Now just one more ball game to complete; the task would then be done;
Just three more aircraft to get off, beneath the broiling sun.
The score so far was really great, with no aborts as yet;
Just one more ball game off on time – a record would be set!
The Tower Charlie talked of this, and pride showed in his stance;
As dignitaries gathered round to watch this one big chance.
Now all the eyes were on the ramp, to watch the aircraft flow;
The tension now began to mount – this cell just had to go!
Cherry One was trouble free, and taxied down the ramp,
And Charlie looked triumphant, though his brow was slightly damp!
Then from the watching multitude went up a thunderous yell,
Which bounded from the flight line, and rattled in the dell;
It struck upon each aircraft – rebounded on each screen;
Cherry Cell was at the point – all aircraft in the green!
There was ease in Charlie's manner as he stood up in his place,
There was pride in all his bearing, and a smile shown on his face;
And all the eyes were on him as he checked the minute hand,
And all the tongues were silenced, as he finally took his stand.
Defiance came from Charlie's eyes, and a sneer curled from his lips,
He responded to the accolades with a wiggle of his hips!
The crowd saw how his face grew cold; he sucked in on his paunch;
He gripped the mike with sweaty palm to give the word to launch!
His eyes were glued upon the aircraft, a tension took its toll,
And he calmly instructed Cherry Cell to start the takeoff roll.
Oh, somewhere in this favored land the sun is shining bright.
The band is playing somewhere, and somewhere hearts are light;
And somewhere men are laughing, and somewhere playing sports,
But there is no joy on the Isle of Guam – Cherry Cell had two aborts!!

A Baggage Drill To End All Baggage Drills!

Derek H. Detjen

November 24, 1968, was a day we'd been looking forward to for a long time. We were scheduled for a final redeployment from Kadena AFB, Okinawa to Guam just prior to returning to the states at last! Although it was the end of our third tour in Southeast Asia, we had loaded up with all kinds of stereo, china and other stuff at the Marine NAS Futema, and other BXs and off-base stores. The weather was overcast and looking ominous as we, along with our sister crew for that afternoon's mission, began to load all of our accumulated luggage onto the bus in front of the BOQ.

The first stop on our way to the pre-mission briefing was the inflight kitchen to pick up the lunches, coffee, and water jugs. To our surprise, the bus driver had jammed the transmission into first gear and could not disengage it. Driving to tactical communications and then to the briefing room, we were already 15 minutes late, much to the consternation of "the colonels." While we were attending the briefing we were informed that the powers that be had ascertained that it was a "safety violation" to drive the bus onto the flight line in its present state.

It was raining softly as we exited the briefing, proceeding to unload and re-load our luggage onto a new bus, some of it through the bus windows. By the time we arrived at the aircraft we were running short of time. We loaded the aircraft hurriedly, getting soaked while passing extra stuff up into the "47 section." Barely getting done with our preflight in time, we just made engine start time, and yep, you guessed it, number eight wouldn't start, cartridge or otherwise. Despite numerous attempts involving FMS bread trucks and various other experts, the Command Post directed us to "go to the unmanned spare!"

The next baggage drill sequence followed in an even harder rain. We loaded back onto the bus, driving over to the unmanned spare and reloading everything onto the new aircraft. By now the Command Post informed us that we "were now Spare Three" as the rest of the mission had launched uneventfully without us. Waiting for our prescribed "vulnerability time" to run out, we were finally relieved of mission responsibility by the Command Post! Time for another baggage drill!

Could it rain any harder? We doubted it as we once again downloaded all of our stuff, mercifully gaining the relative shelter of the bus with all of our junk, wondering why we'd bought some of it! Well, at least we'd have an uneventful trip were back to the "Q" to try again the following day. Wrong!

Just as we reached the end of the active runway on the perimeter road, the still driving rain and mist got into the bus's distributor and it went flat out dead. A call to the Command Post brought yet another relief bus to our aid and once again we engaged in another baggage drill, passing stuff through the windows and lugging the big boxes one by one into the new bus! Finally, mercifully, we arrived at the BOQ to download all of our load at last. By then we'd have been amenable to conducting a roadside fire sale and selling every last piece of our many treasures!

The following day, we departed Kadena, and as Black 1 (Wave Lead), Maj Rod Busbee, Lt Daniels and crew struck the target and returned to Guam, logging an uneventful nine hours and 15 minutes in the process. A little more than three weeks later, we were all back in the States, and you can bet I told my wife just how much effort we'd expended acquiring all her "stuff," as I'll bet the rest of that unlucky crew did also!

Do any of the rest of you get "testy" when loading the family vehicle prior to a long trip? I know I do, and I think it's still the memories of that long ago baggage drill at Kadena that must trigger my short temper!

The author at the B-52 sextant

Lesson Learned - Never Forgotten!

John D. Mize

After a couple of years as an admin officer, I was assigned to be an OTS recruiter in the upper Midwest. This was just before Vietnam started to heat up and the warmth of Texas was appealing to many in the midst of a Minnesota winter. After about a year of that, I began to believe my own story and applied for Pilot Training. Having hung over the fence at Barksdale AFB as a kid, of course, my first choice was the BUFF.

Fresh out of Castle AFB, California, I arrived at Ellsworth AFB, South Dakota in January of 1969 only to find the 28th Bomb Wing getting ready for an Arc Light deployment. I got my wife settled in base housing, processed into the 77th Bomb Squadron, had a down and dirty standboard ride and was off to Guam as a spare copilot. That began the endless TDY routine and experienced those many moments of boredom, punctuated by sheer terror that we all went through.

As I was sitting in the BOQ in late February of 1969, the phone rang; a copilot had cracked his head open in the O'Club swimming pool and they needed a substitute copilot right then! I grabbed my gear and headed out on my first solo ride as a copilot on a B-52D; I met the crew, attended the briefing and was off to the aircraft.

The weather, as usual, was HOT, HOT, HOT! The preflight was completed and I had all my flight gear on - helmet, gloves, chap kit, water wings. The engines were started; we taxied and were ready for takeoff. I put my parachute on and the check list was completed. The A/C stated:" It's too hot to gear up." He had his headset on, his parachute off and his seat belt latched.

We taxied onto the runway, lined up, powered up, and activated the water switches. At 70 knots the S-1 timing was started; after S-1 and before S-2, the water assist quit and we were at max weight for take-off!

The pilot stopped flying the aircraft to put on his helmet and parachute; I suppose in preparation for disaster. I, the new, very green copilot, assumed control of the aircraft as it departed the end of the runway. At about 100 feet above the water, after a drop of 500 feet, the aircraft gained enough speed to resume its ascent. After level-off, the pilot took back command of the aircraft, accomplished the air refueling, delivered the weapons and landed upon the return to Guam.

On my next flight I drew the same crew. The pilot said, "You make the takeoff from the start." The mission was uneventful.

Lesson learned: As a pilot, your job is to fly the aircraft until it will fly no more.

Time moved ahead to mission number 295, December 27th, 1972. During Linebacker II, I was the pilot in command of Ash 2. As we were inbound to the target, with the PDI centered, at bombs away, 15 SA-2 SAMs were launched directly toward us. 14 of them were evaded. However, number 15 hit the left side of the aircraft, causing the aircraft to invert, taking out four engines on the left side and all the flight instruments. Recovering the aircraft to level flight, we flew from the Hanoi area to NKP, Thailand, using needle, ball, and airspeed indicators and dead reckoning provided by the navigator. As the plane became obviously doomed, the bail-out order was given. All souls ejected safely and lived to tell the story!

Lesson proven: FLY THE AIRCRAFT UNTIL IT WILL FLY NO MORE!

I ended up flying a total of 295 Arc Light missions, the final one being that Linebacker II one on the 27th of December 1972. Since I had accumulated enough "fruit salad" to wear, that seemed to be a good time to go home!

I spent the next 12 years in various flying and maintenance positions and retired in 1984. I was able to put some of my experience to use when I worked as a Console Operator for the B-52 Flight Simulator - kind of like a life size video game! I retired for the last time in 2003 and, with my wife Joan, am enjoying retirement in Belmond, Iowa.

The author in 1973 in a D-model.

Undesirable Emergency

Dwight A. Moore

That's what TIME magazine called it in their 5 June 1972 issue – "Undesirable Emergency." I'm not sure how much of an emergency it was but the liberal press needed a headline and another way to demagogue the military.

The article described the event as: "Even though sovereignty over Okinawa was restored to Japan two weeks ago, the Pacific island continues to be a sticking point in relations between the two countries. American military installations are to remain on the island for the indefinite future. The Japanese are now concerned that Okinawa could involve them, even indirectly, in a war. A case in point occurred only days after Okinawa ceremonially changed hands. Three Guam-based B-52s were unable to refuel in mid-air on a bombing run to Vietnam because of weather conditions in the western Pacific. They were diverted to Okinawa's Kadena AFB, where the big bombers were based until last year. Aware of Japanese sensitivities, the U.S. embassy in Tokyo alerted Foreign Minister Takeo Fukuda about the new flight plan of the B-52s; thus Fukuda was able to break the news of an "unavoidable emergency" that forced the planes to land on Okinawa for a four-hour refueling stop. Nevertheless, a government spokesman agreed with Socialist critics in the Diet last week that frequent emergencies of this kind would be 'undesirable.'"

My observation of the same fact pattern was only slightly different. My crew and I, Carswell E-15, launched from Guam as White 3 in B-52D 56-667 on 20 May 1972 for a routine run in South Vietnam. It was my 59th combat mission in the B-52D, but only my 9th as an aircraft commander, so we were still getting some easy runs. The flight was uneventful until we caught up with the tankers out of Kadena just north of the Philippine Islands for some nighttime refueling in the Busy Rooster refueling track. There were only two tankers for three bombers. The story from the tankers was that the boom operator had dropped something on the pilot during preflight that had sent him to the hospital so they launched with two and a promise to send us a tanker on the way back to Guam. I took on 58K pounds of fuel and White cell headed for SVN. The bomb run was uneventful and White 1 was on the radios prior to feet wet trying to learn our new game plan.

We were supposed to contact a tanker who had launched out of Kadena on what was his first mission since deploying to the Pacific. The tanker didn't respond to his tactical call sign - the only thing we had - and was looking for some aircraft with tactical call signs that didn't match ours. Plus, he apparently had never heard of White cell. By the time we collectively realized that he was for us, he was roughly 100 miles behind and we weren't going to make it.

8th Air Force had carried Kadena as an alternate landing field in case of emergency for some time and was eager to see if that was still a good option now that the island had reverted. I had actually been on Kadena the night of the reversion expecting a celebration by the locals, but was surprised at how totally dead the island was - only the taxi cab drivers were out and they were grumbling about the Japanese tax on taxi fares that started that day. As 8th AF now had an opportunity to test out the use of Kadena and we had an opportunity to get some fuel before we ran out, the diversion made eminent sense.

I landed last and as we were taxing off the runway the cars were already starting to line up along the perimeter of the base. Having been a captain for a whole six months, I was pleased to be greeted by an O-6 as I departed the plane - must not get many visitors. Unfortunately, he was there to tell me I had four hours to get my plane and my sorry butt off of the island. That was at about the same time as the copilot pointed out that the front tire had been chewed up pretty badly either on takeoff or on landing and a lot of the cord was showing through. It needed to be changed. The colonel wasn't amused since it would take at least half a day to get a tire flown in from Guam along with someone who could change it. We would then be out of crew duty day and the locals would really be upset and probably rioting by then. That was not an option. Having spent 11 hours flying to get to this point, I really needed sleep and a beer, but I told the crew to get some fuel for the three-hour flight back and a load of water. I left the copilot and gunner in charge and went to mission plan.

True to his word, the wing commander wanted us off in four hours and had us fueled and at the end of the runway before the time expired. With a light load of fuel and a full load of water for takeoff, I was determined to get my aircraft off the ground as quick as possible. The Dash 1 doesn't recommend lightweight takeoffs with water, but they sure are fun. After one and two did their normal roll down the runway, we taxied out. It was great acceleration for a D-model and a short roll till we were airborne. Leaving nice thick clouds of black smoke, we did a maximum angle takeoff climb over the field. When everyone has come to see it, you might as well put on a show.

Anyway, it was an uneventful three hour trip back to Guam, but one really long day for the crew.

"Your Target for Tonight"

Bruce Woody

When I was presented the opportunity to write for this wonderful series of books I was very much undecided as to exactly what I wanted to write about. I got into Arc Light a little late; my crew arrived at Andersen the night of the Watergate Burglary in June of 1972. We had a recently-upgraded pilot, Dale Hammerlund, with one tour behind him; Bill Lamb, a very experienced copilot also with a tour; Bob Kuzma, a Radar who had B-47 time and had just come out of C-130s; an EW I had gone to Nav School with, Ron Maxey, on his first tour; and Jud Howard the Gunner (the old man—I think he was 40!) who had just transferred from teaching at gunner's school at Castle. Jud had trained probably 75% of the gunners then in SAC and knew everybody. And there was me, a freshly-minted Nav with 300 hours in definitely non-threat areas, mostly "bombing" Lake Mead in a T-29. I had finished BUFF transition at Castle and was ready to go to war. It was a good mix, and a darn good crew as it turned out. We flew together for a year and were, of course, "The Very Finest Crew in SAC." We were R-26 from Dyess AFB, Texas, for anyone who might care.

I thought about the very first mission we flew, when everybody has their survival vest on and the trusty .38 snub nose all primed for our mission over the Delta. I thought about writing about the transition to U-Tapao after a couple of months and the wonder of seeing Asia for the first time, and how the smell of garlic never left you the whole time you were there. Or the time a refueling manifold broke during preflight and we sloshed through fuel getting out of there. I could have written about when we had to write up the IN chute because of an unfortunate occurrence with the relief bucket. There were also the stories about our Guam Bomb, which broke apart going around a curve, probably because when one sat in the back seat one could look down and see the road going by. You know - regular stuff like that?

However, after due deliberation, I decided to write about our crew's experience during the Eleven Day War, a/k/a Linebacker II.

We had just returned from our first 28-day break, during which we were able to catch up on the news, including Henry K's famous "peace is at hand" remark. Then we saw President Nixon re-elected and wondered if we would go back after all. We did go back and flew routine missions for a couple of weeks. It was the evening of December 17, 1972, at Andersen AFB, Guam and we were eating dinner at the O-Club, which was a rarity for us. An announcement was made that all crew and support personnel were restricted to the base. That's it; no explanation was forthcoming. I said, "Well, we're either going home or going to Hanoi."

We finally did get word that we were to have a mass mission briefing in the early afternoon, which still could have meant either alternative, although we felt more likely we would be going west rather than east. Everyone gathered in the large briefing room in Bomber Operations and the briefing officer approached the overhead projector. "Gentlemen," he intoned, "your target for tonight..." He then uncovered the screen and there was a big triangle in the middle of Hanoi. The room was full of the starkest silence I have ever not heard for about 10 seconds, and then the briefer tried to take the edge off with a joke: "Your support will be one Navy F-111." That did the trick, all right. We didn't know whether to laugh or cry, but we did talk. We weren't totally sure he was kidding.

I think there were around 80 sorties from Andersen that first night, and we all broke off for our specialized briefings. When we got to our BUFF, we did a pre-flight that would have made Standboard proud, especially the ejection seat part. Stand-by crews had already done the engine-off preflight, but I'm afraid their efforts were wasted, especially the ejection seat part. I won't attempt to describe our feelings; most of us tried our best to put on a routine face. It didn't help when, at the mass briefing, it was announced that clergy was available for those who wanted a little counseling. The local Catholic chaplain had a little altar set up in a small briefing room where he was hearing confession. I'm not sure, but I think one Southern Baptist crewmember, which will remain nameless, may have gone. Even I, a good Methodist, thought about it.

Takeoff and climb out were as routine as they could be, considering about 80 BUFFs were taking off with either a minute or minute and a half separation. Naturally, we tested everything on the aircraft during the long flight over. I even paid attention to the astrotracker, which was essentially useless during the routine missions. I wanted everything I could get up and working since this was going to be a synchronous run, which we rarely did over there, and I wanted the best heading information I could get into the Bomb/Nav system.

Well, there was a glitch. We were flying at night, and the filter which popped in automatically to protect the optics when the sun was shining wouldn't pop out. The astrotracker thought it was daylight, so it couldn't lock onto the stars. I tried everything: circuit breakers, hammer, questionable language, prayer, you name it - all useless, to my immense frustration. Then I got really desperate, and I got to thinking: "It's probably an electrical short somewhere and maybe I can bypass it." So, totally without logic, I broke out the amber light that was on, which was indicating that the sun filter was in. I stuck a piece of aluminum from someone's pack of cigarettes into the hole where the bulb had been, and, lo and behold, the filter went down and the thing locked onto a star! I felt very proud of myself, although when I reported this to maintenance guys after the mission, all I got was a dirty look from the five-striper from OMS.

At any rate, in we went about halfway between Saigon and the DMZ. Obviously course and timing were doubly important with all the BUFFs in the air that night which included a bunch from U-T. Somewhere in the area of Udorn, we cut northeast to approach Hanoi from the North. This necessitated flying almost to the Chinese border and then cutting south to the Big H. Picture, if you will, a string of over 100 B-52s heading straight for China. Picture it from the viewpoint of Chinese early warning radar operators. When we were about 100 miles short of China, the EW called out, "The Chinese are looking at us." When we were about 50 miles short of China, he called out, "Now they are REALLY looking at us." To their immense relief, I am sure, we turned south.

IP inbound. Correct the course; what's the drift up here? Synchronous checklist. Recheck offsets; don't be the one who bombed offset-direct and took out the orphanage. SAM calls all over Guard. Beepers all over Guard. SAM calls and beepers all over each other. Red Crown constantly calling MiGs from "Bullseye." God bless the Navy. Navigation lights out. Geez, Louise. Pilots can't maneuver inbound; might dump the bombing gyro and/or run into somebody. Four sets of eyes on the altimeters. Straight in. Gunner calls SAM. Concentrate. Gunner calls another SAM. Beep, beep, beep. "Red Crown has target 240 at 75, Bullseye." Beep, beep, beep, beep, beep. Copilot calls SAM. Cross hairs over correct bend in river. Offset one in. Looks good, Radar. Gunner calls two SAMS. Offset two, looks good. New levels of fear being explored. Job getting done, despite being scared out of one's wits.

Copilot: "Boy, that one was close."

Pilot: "Yeah, I could read the Russian letters on it."

60 - TG. Stopwatch going. Offset is in. Beep, beep, beep. 30 - TG; radio saturated; good offset; RCD connected, light on; Bric Select, light on, dim. Bombing system switch, Auto. Etc., etc., etc. SAM, SAM, SAM. Beep, Beep, Beep. Bombs away. EAR. Turn, turn, turn.

Nav to Pilot: "Maneuvers authorized."

Pilot to Nav: "I AM maneuvering." (One had to be reminded in the BUFF.)

Slight rise in my seat. "Don't get carried away, Pilot," I say to myself.

N-1 inching over. On heading, more or less, but close enough at that point. Watching doppler groundspeed spool down as we enter jet stream 150 knots slower than when we came in.

"Who planned this mess?"

Nav: "Crew, we've just left the last SAM ring on the chart."

Gunner (immediately thereafter): "SAM, 6 o'clock." Note for intel debriefing. Foil falls out of Astrotracker.

Nav: "Just crossed Thai border, crew." We immediately lose 10 knots with the collective sigh of relief.

"Red Crown, Rose is out with three."

Time to relax and think in complete sentences again.

The plan was to proceed back to Guam as usual. However, we were informed by HF that there has been a major mission planning malfunction, and somebody on the ground at Andersen had miscomputed the fuel curve, and we were several thousand pounds short.

We took it in stride; our perspective had been changed a little. So, we had to fly northeast toward Kadena to meet tankers. OK, fine. However, we were governed by our understanding with Japan, which had just taken Okinawa over again. If we had hangers, we couldn't land at Kadena; we would have to ditch. After discussing for a short while what the Japanese could do with their little rule (even though we had no hangers), we decided we would have landed there anyway, if we had to. We would rather be "courted" than "ditched."

Approaching Kadena, the Radar and I noticed several large returns over the water. That's right – weather. Thunderstorms, bad weather, you get the idea. We bumped along; Pilot got the gas; we came back. It had lasted 18 and one-half hours. Two more missions to go before it was over, but they would be from U-T. On the last one, the Bad Guys didn't have much left to throw at us.

Several things became crystal clear. He who is not scared out of his wits in such a situation is too crazy to fly. And, there is a good reason for all the training and repetition: When you're too scared to think, you can do the job anyway.

It was a good job, I think. The operation broke the logjam in the peace negotiations and resulted, we believe, in the release of our POWs shortly afterward. We were proud to have gotten the POWs released and restored to their families, and that made it worth it. We did reflect that since the last missions of the operation were essentially non-opposed, the Bad Guys being out of SAMs and the fighters staying away, we could have won the whole thing with a couple of Marine battalions on the ground.

But, there was nothing we could do about it, so on to the next show.

Gunner Traditions - Alive and Well

John R. Cate

The following is dedicated to Beau Howard, my friend and comrade of many years, and to Don Murphey, who I wish I had gotten to know sooner. Without their help this story wouldn't be possible.
Thanks Bo, thanks Murph.

Nining M. Guzman, Mama Nining or just plain "Mama." Three common names for one very uncommon woman. She had an easy smile and a spontaneous laugh. She was only 5' tall and weighted about 100 pounds soaking wet, but was blessed with a beautiful soul and a giving heart. For such a small person, she was to have a great and lasting impact on B-52 gunners, giving us one of the most important traditions we have today. I had the honor of being the last Wing Gunner in the 43rd Bomb Wing, serving in that position from 1988-1990. During those two years I came to know Mama Nining very well and was privileged to call her friend. I spent many enjoyable hours with her. This story is being told so that her 26 years of caring and involvement with all B-52 gunners will not be forgotten.

Mama was born Antonina Muna Guzman on March 1, 1919 to Tomas and Ana Muna Guzman on the island of Guam. The Guzman family was a well known and well respected Chamorro family. She had three sisters and four brothers. Two of her brothers, Jesus and Jose were serving with the Insular Force Guard, a local defense unit assigned to the U.S. Navy at the beginning of World War II. On

Monday, December 8, 1941 nine aircraft of the Japanese Imperial Air Force based on the island of Saipan, bombed the village of Sumay, then the island's commercial center and later strafed the villages of Piti and Agana. It was Sunday, December 7, 1941 in the United States. Four hours earlier, Pearl Harbor had been attacked by the Japanese Imperial Navy.

The people of Guam were preparing to celebrate the feast of the Immaculate Conception. By the end of the day, the feast would be transformed into the beginning of one of the most tragic periods in the history of Guam. For Mama and countless millions of others around the world, it was to be a day that would change their lives forever. Wednesday, December 10, 1941 dawned clear and the sea was calm. That morning over 100 Japanese naval ships flying the flag of "The Rising Sun" approached the island. Soon a force of 400 members of a special navy land force and about 5,500 army troops invaded Guam. History records that at 0700 hrs Capt. G.J. McMillin (USN), Governor of Guam and Commandant of the Naval Station, Guam, facing overwhelming odds and certain defeat, surrendered the island to the Japanese. The 400 members of the special naval land force came ashore on Tamuning's Dungca Beach. This force, having regrouped, moved inland, making their way to Agana. On the way they ran into a group of Chamorro families, including Mama's, fleeing the area in a small bus. The Japanese troops opened fire on the bus. Those not killed instantly were bayoneted; 13 men, women and children perished. Mama's parents and brother, Edward were killed outright, Mama was bayoneted in the stomach. In the confusion, Mama's three sisters and brother Gregorio or Greg as he is known slipped away into the jungle, unharmed.

Hiding in the jungle, was Ricardo Bodallo, who ran forward and dragged her away before she could be bayoneted a second time. His unselfish act of bravery saved Mama's life. Ricardo was only 14 years old at the time of this tragedy. Mr. Bodallo, who later served two terms as Governor of Guam (1975-1979 and 1983- 1987), treated Mama's badly infected wounds and slowly, over time, nursed her back to health. The night of Edward and her parents' killing, Greg joined a small Chamorro resistance force and would harass the Japanese until 1944. He, along with other members of the resistance lived in the caves around Talofofo Falls.

Talofofo Falls has one of only two accessible fresh water supplies on the island and was well hidden from the Japanese by dense jungle growth. During this time, Greg would "requisition" any medicine Mama needed for her recovery from the main Japanese Army Hospital located in Agana. Under the cover of darkness, he would bring his sister this medicine, along with any news of the resistance movement. He did this many times over the next two years, never getting caught, but as Mama would later say, "Too many close calls". Greg would survive World War II and with other members of the resistance, would fight along side the U.S. Marines until the battle of Guam was won and the island was declared secure on August 10, 1944.

Mama Nining

After the war and with life returning to normal, Mama and her sister, Josefina started a dry goods business in the village of Mongmong, about five miles outside the front gate of Andersen Air Force Base. This store proved to be Guam's first successful department store and would continue to grow and prosper for the next 20 years. In the late 1950's, early 1960's, Mama and Josefina had a falling out. Out of love and respect for her sister and having a strong sense of family, Mama gave her half of the business to Josefina. This store was later acquired by the Liberty House department store chain. Around that time Mama started her first store, Mama Nining's Friendly Tavern. It was located just behind the dry goods store. She sold hot dogs and cold drinks to the island's growing construction force. During that time, Guam was in the middle of a major construction and expansion period, both at the US Naval Base and Andersen Air Force Base.

Official US Air Force history records June 18, 1965 as the date that Guam based B-52's made their first strikes against targets in South Vietnam. Operation Arc Light had begun. B-52 gunners remember Friday, June 18, 1965 as the day the air war against North Vietnam began and it wouldn't end until Wednesday, August 15, 1973 when the last strikes were flown against communist targets in Cambodia.

During those eight years B-52 gunners naturally found their way to Mama Nining's Friendly Tavern while TDY to Guam. It started with just one gunner, Paul Kiviaho, TDY from Ellsworth Air Force Base. In talking with Mama I found out that sometime in 1965, Paul found her place by accident. Having a down day, he borrowed a Guam Bomb (old, rusty car) and was exploring the island. Hungry, he saw Mama Nining's Friendly Tavern and stopped. Mama and Paul seemed to hit it off right away. He acquired a taste for her hot dogs and cold beer. Soon, Paul was a regular patron at Mama's.

According to Mama, Paul was the first gunner to receive a bean from her. I know that several of my fellow gunners may disagree with this, but that's how Mama remembered it. I once asked Mama why she gave Paul a bean and she answered, "Mama's bean good luck". Mama went on to tell me that the bean was an old Chamorro custom. She than swore me to secrecy and told me her "beans" came from a small grove of Philippine Sea Trees just south of Tarague Beach on Andersen Air Force Base. Her brother Greg, now a professional fisherman, would collect the beans for Mama when he was fishing the leeward side of the island. Many don't know this, but Greg is the fisherman who caught a world record sailfish, fishing the Marianas Trench off Guam. Today this magnificent fish is mounted and hangs in the arrival area of the Guam International Airport.

Mama kept her supply of beans in a small paper bag in the cash drawer of her register. Only a handful of gunners ever knew where Mama stored her beans. When it came to issuing a bean, Mama always followed her checklist. First, a gunner had to "qualify" for a bean. If Mama felt you were just stopping by to get a bean, if you felt you automatically qualified for a bean just because you were a gunner, if you didn't want to spent time with Mama and your fellow gunners, if you got drunk and were loud and obnoxious and she had to ask you to leave she wouldn't issue you a bean.

Once Mama decided to issue a gunner a bean she would take one from the paper bag and rub it on both sides of her nose. She said she rubbed the bean on her nose for good luck. Next, Mama taking your right hand, turning it palm up, would put the bean in your hand and closing your fingers around it, would look you in the eye and say "As long as you carry Mama's bean no harm come to you". Finally, Mama taking your face in her hands, would kiss you. In later years Mama would ask the gunner to sign a ledger or "my guest book", as she called it. Unfortunately, these ledgers were lost in a typhoon. During my two years on the island, I personally witnessed this ritual at least two dozen times and she never once deviated from this procedure.

Paul told his fellow gunners about Mama's and soon, two and three, sometimes four or five gunners at a time would make the nightly pilgrimage to Mama's. Most remember the next gunners to receive their beans were Bill Whitenmier, Bob Netzger, Bob Herring, Dale Anderson, Jim Siendenburg and Don Murphey. All gunners, no matter at what base they were stationed, were always made to feel welcome by Mama. The gunners I have personally had the opportunity to speak with remember Mama's Friendly Tavern as their home away from home. She loved to cook and entertain for all "her" gunners. My personal favorites were Mama's Spam Fried Rice and Booney Pepper Popcorn. "Booney" was the island term used to describe just about any and everything on Guam such as booney car, booney house, booney dog or cat. Actually "booney" is island slang for jungle. Booney peppers were peppers that grew wild in the jungle. They were about an inch long, could be found in a rainbow of colors and were extremely hot! Mama would crush these peppers up and mix them with the popcorn oil as her popcorn popped. When eating booney pepper popcorn, ice cold beer was a necessity. Mama's soon became a "first stop" destination for both deployed and PCS gunners and would remain so for the next 26 years.

Sometime in 1970, Don Murphey started the practice of giving gunners a gunner coin. At the time Don was the S-03 Gunner at March Air Force Base. In the past he had given items such as bulldog ashtrays to gunners going PCS from March AFB. In the process he had become friends with the Sales Manager at the Mack Truck dealership located outside March AFB. One afternoon while purchasing several ashtrays, Don was given a small paper bag by his friend and asked to look inside. He found 12 bulldog coins. At the time these coins were given to any Mack Truck salesman who attended the annual Mack Truck Sales Meeting in Allentown, Pennsylvania. The coin featured the likeness of a bulldog on one side of the coin and the words "You Made The Difference" engraved on the reverse side. The bulldog on the Mack

Truck coin bore a remarkable resemblance to the SAC bulldog, the unofficial symbol of B-52 gunners. Later the phrase, "You Made The Difference" would be changed to "You Make A Difference" and a date such as "1972" would be stamped on the coin just above the bulldog's back and rump.

Don kept one of these coins for himself and gave the other 11 to his fellow gunners. This gesture was well received by his fellow gunners. Other gunners begin asking the March AFB gunners for a bulldog coin. In response to these requests, Don ordered 100 more bulldog coins from Mack Trucks, Inc. All told, Don reckons he ordered 300 bulldog coins. During this time, just as other B-52D model gunners were doing, March AFB gunners were pulling back-to-back TDY's in support of Operation Arc Light. Several of Don's fellow gunners always seemed to have a supply of coins with them and would give one to any gunner who requested one. Several gunners I have spoken with remember Don giving them a gunner coin in the NCO Club while TDY at U-Tapao Royal Air Base, Thailand. By then word had spread about Mama's bean and the gunner coin and as a result, the beginnings of the B-52 gunner coin/bean tradition were taking hold.

In late 1973 with the air war against the North Vietnamese coming to a close, the 90-day TDY rotations to South East Asia (SEA) and Guam were starting to wind down. A combination of Mama's bean and the gunner coin had given B-52 gunners a sense of identity and true camaraderie. As gunners rotated back to their individual bases, gunner associations were being formed. Each association had duly elected officers to include a president, vice-president, secretary and treasurer. Most, if not all associations contributed to local charities such as the Make-A-Wish Foundation and participated in local events such as chili cook-offs and base open houses. Fourth of July, Labor Day and Christmas parties were held every year and soon became annual events.

In the late 1970's (the exact year is not known), the practice of issuing new gunners a gunner coin along with a copy of *Rules of the Gunner Coins (Bean)* was begun. To earn the right to call himself a gunner and carry a gunner coin the gunner candidate had to successfully complete gunnery training. At the time B-52G and H model training was conducted at Castle AFB, California and B-52D model training was conducted at Carswell AFB, Texas. To successfully complete his training, the gunner candidate had to first graduate from ground school and then flight training. When the gunner candidate graduated from ground school, he was awarded a set of Basic

Enlisted Aircrew Wings. In later years, the Basic Enlisted Aircrew Wings would be changed back to the World War II era Gunner Wings. To complete his training the gunner candidate had to satisfy all the requirements of flight training and pass his initial checkride. The next duty day after this checkride, the gunner candidate, along with his flight instructor and Stan/Eval evaluator would meet to debrief his checkride. Also in attendance was the Wing Gunner or his assistant. Once the debrief was complete and it was known that the gunner had passed his checkride, the Wing Gunner presented a gunner coin and a copy of Rules of the Gunners Coin/Bean to the new gunner.

Mama's Friendly Tavern

In 1985, Marine Drive was widened and relocated as it passed through Mongmong. Mama lost her store's location to this road construction. GovGuam compensated Mama by giving her the road construction crew's compound. It consisted of a building with 10 bedrooms, two bathrooms and a small building that had been used as the company's office. Several 60th BMS gunners including Beau Howard and John Wing pitched in to help Mama remodel the main building of the compound into a tavern. Beau, John, and a group of US Navy Seabees, including Ed Hord salvaged lumber, paneling and anything else Mama could use from the old tavern. They built a bar, moved her beer coolers and supplies and did whatever else was necessary to help Mama reopen as soon as possible. Later the Seabees built a barbeque grill and poured a concrete slab that served as a patio. Ed managed to secure several large wood cable spools that Mama used as tables. A gunner donated a B-52 drag chute that was used as an awning over the entire outdoor area. Beau and John cut the grass, put up new signage and rehung Mama's collection of pictures and plaques. This was Mama's second and last tavern and was frequented by gunners until October 1991 when gunners were eliminated from the B-52. It was later destroyed by a typhoon in 1996 and was never rebuilt. There were no gunners left to help Mama rebuild.

Mama's Friendly Tavern not only served as destination for gunners, but also enjoyed a large clientele of Seabees as well. Gunner parties and cookouts at Mama's were well known and are fondly remembered and were always open to the Seabees and the Seabees would always reciprocate, extending an invitation to the gunners for any of their functions. One Seabee couple who were regulars at Mama's was Ed and Sherrie Hord. Ed was a retired Senior Chief and Sherrie was an active duty Master Chief. Both were good friends of Mama. Louie LeBlanc was still alive and would visit Mama's at least once a week. Louie would later die of a heart attack while stopped at a traffic light on Marine Drive.

When visiting Mama's, many may have seen Gene and wondered who he was. Gene was a regular fixture at Mama's. He had come to the island in the early1960's to work for a construction firm. After only working for several weeks, Gene fell from a scaffold two stories up and suffered a severe brain injury. He would recover at the Navy Hospital and was allowed to return to work. On his first day back, Gene was ask by his foreman and a safety inspector to demonstrate how he had fallen. Gene climbed the same scaffold, slipped and fell off again, suffering another brain injury! I know this may be hard to believe, but it really happened. This time he wouldn't fully recover. One day he wandered into Mama's and never left. He lived at the bar and Mama cooked for him and did his laundry. In return, Gene cleaned up the bar and tried to help out in anyway he could. Everyday Mama would give him a few dollars in a small paper bag, along with a can of Pepsi. Gene would walk the island and return every night around sunset. On the Wednesday before Thanksgiving, 1988, Gene left Mama's and was never seen again. Beau and John took Gene Thanksgiving dinner and he wasn't there. We searched for Gene for several weeks, but never found him. Mama was heartbroken.

The year is 1990. For nearly a month rumors fueled by stories in the Guam Pacific Daily News and the Stars and Stripes, had been circulating about the deactivation of the 60th Bomb Squadron and B-52's leaving the island for good. The night before the announcement was to be officially made, the 43rd Bomb Wing Commander, Colonel Hall called me to his office and gave me the word. Later that evening, I told Mama the rumors were true and we would be leaving Guam in a few months. Mama started crying and said "Mama knew rumors true". We met with the 15th AF commander later that month at the Andersen AFB NCO Club and were told "you'll be taken care of", i.e. getting the assignment of your choice. Most did.

Looking back, I know my assignment to Guam was the highlight of my gunnery career. I had an outstanding staff: MSgt Tom Lindsey / Tactics Gunner, MSgt John Turner / Squadron Gunner, MSgt Joe Robertson / S01 Gunner, SSgt Beau Howard / Training Flight Gunner. Fellow gunners John Wing, Doug Lunsford, Bill Johnson, Paul Hartman, Steve Gramling, Shawn Daughtery and others whose names I've forgotten, but whose faces I can still see, served with pride and professionalism and would have made the Arc Light gunners who came before them, proud. We had one last gunner party at Mama's and told her goodbye. The next day I left the island, stopping by Mama's one final time. She kissed me on the cheek and reminded me to always carry my bean. I never saw her again. Mama died on March 24, 1998. She was 79 years old. She is missed by all those who knew and loved her.

Over the years Mama's bean and the gunner coin evolved separately and yet both came to identify a B-52 gunner. Today, most B-52 gunners think of the coin and the bean as one in the same. We also know that the gunner coin will never replace the bean and it was never meant to. Mama's bean will always belong to B-52 gunners. What must be remembered is today the gunner coin is no longer uniquely B-52, but belongs to all aerial gunners from World War II through the end of the Cold War. The gunner coin unites us all. It serves to remind us that we are truly a "Band of Brothers". We have all known the joy of flying the "Thing of Beauty" we love, whether it was a B-17 or a B-52. We all have shared similar experiences, have known similar fears, have similar "war" stories to tell to our grandchildren. Most of us have experienced combat but some have not. Many of us have known the boredom of years of "setting alert" during the Cold War. We have all lost friends in the line of duty and in defense of our great country. In the end, we are bound together by those experiences, having found our common ground, discovering a true sense of camaraderie. Because of this realization and understanding, the Air Force Gunners Association (AFGA) was founded in 1986 by James F. Zaengle, William F. Dalton and Fredrick G. Arthur. From its humble beginnings of only three members the AFGA today has approximately 1,325 members and continues to enjoy a steady growth. Membership includes gunners from World War II, the Korean War, the Vietnam War and the Cold War eras. The only requirement for membership in the AFGA is that you were once a gunner. Today, when a gunner joins the AFGA he is issued a gunner coin along with his membership card. Because of the care and concern shown by Mama Nining for "her gunners" and through the efforts and foresight of gunners such as Don Murphery and many others, gunner traditions are truly alive and well. *C'est La Vie.*

Rules of the Gunner Coins (Bean)

1. Bulldog coin and bean are equal and interchangeable, i.e. having a bean is like having a coin. Having a coin is like having a bean.

2. Coin or bean must be carried at all times. You can be challenged for it anywhere anytime. You must be able to produce the coin or bean without taking more than three or four steps.

3. When challenging, the challenger must state whether or not it's for a single drink or a round of drinks.

4. Under no circumstances can a coin or bean be handed to someone. If a gunner gives his coin to another gunner, that gunner can keep the coin – it's his. However if a gunner places the coin down and another gunner picks it up to examine it, that is not considered "giving"

and the other gunner is honor bound to place the coin back where he got it. He cannot challenge while he holds the other gunner's coin.

5. If a gunner has never been given a coin or bean he cannot be expected to play the game.

6. Rules of the game must be explained to all new coin holders.

7. Lost coins or beans, or failure to produce said coin, results in the challenger being bought a drink or a round of drinks. This type of transaction could be expensive, so hold onto your coins. Once the challenged has bought you a drink you cannot continue to challenge him.

8. If a coin is lost, replacement cost is up to the individual's own gunner association. A lost coin should be replaced at the earliest possible time. Losing a coin and not replacing it does not relieve a gunner of his responsibilities.

9. Gunner coins should be controlled at all times. Giving them to anyone is like opening up the various associations to anyone. It's up to the individual associations as to who, outside the active members, they want to have the gunner coin or bean. It's considered an honor to be given a coin and let's try to keep it that way.

10. Local base rules apply to local gunners only.

11. The above rules pertain and apply to anyone who is worthy to hold the position of a defensive aerial gunner, (or) has held the position or has been selected as an honorary member by an active association.

12. The coin will not be defaced – i.e. drilling a hole in the coin.

The author and crew with Brig Gen Jimmy Stewart.

Our Flight with the Star

Robert C. Amos

Crew E-17 of the 454th Bomb Wing from Columbus AFB, Mississippi, was TDY to Andersen AFB Guam as part of the 716th BS/ 420th BS Composite Wing in 1966 conducting Arc Light missions over South Vietnam (SVN). It was our second 89-day TDY tour and we had already accomplished 20 combat missions over SVN starting with the second Arc Light mission on 5 July 65.

I was looking at the flight schedule early on 20 February as we began flight planning for the sortie to be flown the next day. A Brig Gen Stewart was listed as an extra Pilot. I hurriedly asked our Squadron Commander, Lt Col Collins Mitchell who was this General Stewart? He replied that, "You know Bob, its BG Stewart, the Actor - he's over here on an active duty reserve tour, and we wanted him to fly with a young crew, reminiscent of his WWII crews he commanded in England". With this knowledge, we then settled down to some serious mission planning and prepared a series of extra maps and charts depicting the mission over SVN.

It was a somewhat simple sortie where we were to bomb a suspected Viet Cong stronghold and bivouac area NW of Saigon where the only the potential threat were the Cambodian based MiGs. We were Green 2 in a 30-ship bomber stream named New Car I. General Stewart and our Wing Commander, Col William Cumiskey arrived at the mission briefing late that afternoon where we outlined the mission, air refueling, and recovery procedures back at Guam. General Stewart would fly in the IP's seat directly behind the pilots during the mission. We immediately had the Gunner, TSgt Demp Johnson, go to the Andersen Commissary and buy fresh eggs, bacon, and bread so that we could fix scrambled eggs and bacon along with grilled cheese sandwiches on the way back to Guam.

When it came time to contact the Tanker Boom Operator while in the pre-contact position behind the tanker, I had General Stewart repeat the "Green 2 stabilized in the pre-contact position, ready for contact." There was an unusually long pause before the Boom Operator responded. After plug in, General Stewart replied "Contact." Later as we completed refueling, I informed our tanker of our distinguished extra Pilot

As we approached the coast of SVN, and were running our checklists, General Stewart asked if we were at the Pre-IP point on the map. He then moved to the edge of the IP seat so to view the bomb impacts of the aircraft ahead of us. All went well throughout the bomb run as the RN, Captain Irby Terrell found and tracked the bombing offset, and the TG started the countdown. At TG zero, the BRIC started the release of our 51 M-117 bombs.

When safely outbound of SVN heading east, we pressurized the cockpit to 7.45 PSI from the combat setting of 4.5 PSI, and settled down for the flight back to Guam. Capt Kenny Rahn, the EWO, plugged in the electric frying pan and prepared the scrambled eggs and grilled cheese sandwiches. A wonderful breakfast was had by all. Of course, our tail Gunner had to bring his lunch, as in the B-52F, he was some 100 feet behind the forward crew compartment.

On arrival at the descent point on the Andersen TACAN fix, Green Cell started descent to intercept the ILS Localizer and Glide Slope. Slowing to 180 knots, I called for the Flaps. Capt Lee Meyers called out the flaps were starting down - then suddenly he called out "The flaps are splitting." I answered "Where?" He answered "Flaps 20," then I said "Bring them up." At the same time, we experienced a rolling moment to the left. The Gunner tried to view the flaps, but they were full up by the time that he positioned himself to view them.

I immediately pulled out of the bomber stream and declared an emergency while climbing to FL200. This started the SAC Command Post into action. A flaps-up landing in the B-52 is possible, however it is only demonstrated in training down to an altitude of 500 ft above the runway, and no actual touchdowns are made. We entered the abort area north of Andersen, and started to calculate the flaps up landing data: airspeed +35 knots; landing roll – longer. Even longer for drag chute failure - 50% longer. Then Major General Crumm, 3rd Air Division Commander, came on the radio asking if we could verify that the flaps had actually split. I responded that there was a rolling moment to the left, but not severe and could have been from the B-52 ahead of us. He then asked if we would move General Stewart down to the Instructor Navigator position in case of a possible bailout and then try flap extension again. I looked over at the Copilot, who shrugged his shoulders and nodded OK, at the same time the rest of the crew called out "We are with you Pilot." I then notified the Command Post that I was going to extend the flaps again. General Stewart was in the IP seat intently listening as I reviewed the bailout order in case I lost control of the aircraft. I then escorted General Stewart to the lower deck and helped him adjust his parachute. I told him that if I lost control of the aircraft, I would callout over the interphone "Bailout" three times and activate the Bailout Light, whereby the Navigator would be the first to go, creating a large hole by his exiting ejection seat that General Stewart was to use it to bail out of the aircraft. I assured him that I would do everything I could to regain control of the aircraft and that I would be the last to leave, but that he was to be the second following the Navigator. In his familiar granular voice he said "Yes, Captain Amos, I understand." I could envision the newspapers if tragedy befell us with - *Jimmy Stewart Killed in Bomber Accident - Amos Pilot*.

Captain Meyers then placed the Flap Lever down, the gauge indicated a splitting condition, but the flaps extended normally without any rolling moments. It was a bad flap gauge, not a flap malfunction, and our previous rolling moment was most probably from turbulence of the bomber ahead of us in the stream. Upon notifying the Command Post of our success, General Crumm told us "To bring her in." To the dismay of the other 29 crews who had landed before us, we popped over the horizon heading westerly to the Andersen runway with the flaps down. There was to be no flaps up landing at Andersen AFB that day - and everyone could put away their Wiener Sticks.

After taxiing in, all the VIPs met the aircraft, and General Stewart suggested a photo of himself and the flight crew. While we proceeded to the Mission Debriefing area, General Stewart was escorted to the infamous "Beer Barrel" to visit with the flightline and maintenance crews where they enjoyed the swapping of WWII and Arc Light stories for several hours.

The next morning while we were mission planning for the next mission, an announcement came over the loudspeaker that the Wing Commander wanted Captain Amos to meet him immediately in front of the building.

Hurriedly I went outside, and seated in the Wing Commander's car was General Stewart. He thanked me and the crew for his Combat Mission and professionalism during the inflight emergency. He then presented me with a set of personalized autographed photos taken after we landed for each members of the crew, and wished us Good Luck. It was a great experience to have General Jimmy Stewart fly with us. He was truly the same gentleman in person as he had portrayed in his many films.

Inside Charlie Tower.

Blackout in Charlie Tower Or How the War Stopped For a Few Minutes

Dave Hofstadter

In the days of the Strategic Air Command the wing commanders and their deputy commanders were referred to phonetically on the radio. For example, the wing commander (CC) was called Alpha, the vice (CV) as Bravo, the deputy commander for operations (DO) as Charlie, and on down the list of colonels. The DO was routinely in charge of day to day taxiing and flying operations.

Well, during the Southeast Asia conflict, hundreds of B-52s had been deployed to Andersen AFB, Guam by the time of the Linebacker operations in the early 1970s. The DO at Anderson was responsible for such things as directing engine starts, taxiing and launches, landings, coordinating maintenance, ordering mission changes, to name just a few, all on the radio with the crews in the aircraft.

Depending upon the battle situations in Vietnam, Laos and Cambodia thousands of miles away, there could be up to hundreds of aircrews at Andersen needing instructions to start up, get maintenance, land, change planes, taxi, park and so on. To handle this volume of activity, the DO at Andersen moved his control point from the traditional command post in a building, to a tower out by the flight line. And to handle the volume there was a large team of assistant DOs who rotated duty as "Charlie" in the Charlie Tower. They were typically all Lieutenant Colonels and were kept busy constantly, answering calls from the aircrews on the radio.

Now, one day some generals from the Pentagon dropped in on Guam on the way to do some Southeast Asia inspection tours and duty-free shopping. They wanted to see the war. If you wanted to get a feel for the B-52 war - without actually going into combat - you climbed up to Charlie Tower and observed. It was like a high tempo aircraft carrier operations, but with hundreds of massive bombers moving over a hundred acres of ramp onto and off of a single runway.

So here they came, all the local generals—with their colonels-in-waiting, and all the Pentagon generals—with their generals-and-colonels-in waiting, all crowding into Anderson's Charlie Tower.

Now, Charlie Tower was none to spacious, and, naturally all the brass needed to get in so they could hear all the radio traffic and experience the war. So one of the local generals shouted above the radio calls and generals chatting, "Everyone below the rank of colonel will leave the tower!" People left, and there was room in the tower. Now, remember I told you that the Charlie on duty would be a Lieutenant Colonel. That day it was a fiery young Lieutenant Colonel we will name O'Malley. Having already been crowded away from his radio console, O'Malley followed orders and left the tower. But the radio calls kept coming in.

Charlie, this is Blue Lead. Blue Cell is at the hold line ready for immediate departure, over?

Charlie, this is Yellow two. Request expedited maintenance for our HF Radio, over?

Charlie, Green Cell ready to taxi, over?

All the generals smiled and looked at each other and experienced the war.

Charlie, Orange One. Orange Cell is taxiing up to the hold line. Notice Blue Cell is holding there. Please advise.

Charlie, Blue Cell still ready for launch, over?

Charlie, this is Brown One, down, logging 13.6 hours. Request parking spot, over.

It started to get quiet. The general who made the announcement started to look around. Down there, outside the tower, sweat staining his uniform at the armpits, hair disheveled, arms folded, was O'Malley, staring up at the B-52s circling, waiting for permission to land.

Well, the generals experienced the war and left. With a soft harrumph O'Malley was told to get back to work.

But for a few minutes in the war, the ground in Vietnam did not shake; trucks were able to move down the Ho Chi Minh trail; it was a little quieter in Southeast Asia.

But not for long: O'Malley was back in Charlie Tower.

Chapter Five

Bar story [bahr] [stohr-ee] – *noun* - a narration of an incident or a series of events or an example of these that is or may be narrated, as an anecdote, joke, etc. told at a counter or place where beverages, esp. liquors, are served to customers.

The 322d Bomb Sq.'s newest crew, R-35. Standing, from left, are: Capt Robert Stewart Jr, EWO; SSgt. John Bennet, gunner; and Ma. William Clark (pointing to map), aircraft commander. Seated, from left, are: Maj Edward H. Woolsey, navigator, Maj Allen P. Gyving, radar-navigator; and lst Lt Tommie Thompson, copilot.

My Cowboys Story

Tommie Thompson

I was a copilot for three years at Glasgow AFB, Montana before leaving the bombers and trying my luck with fighters. During that time I was on several B-52D crews and had made lots of friends with other Crewdogs. I had been known to play jokes on a lot of different ones as I tried to keep most of them in stitches, but for the most part I was a serious copilot.

Those who play jokes on others often become victims of the same, and I was no exception to this rule. One incident which I remember the most happened just before I left Glasgow for F-4's and Vietnam. I know a certain radar navigator by the name of Jimmie Falk and a copilot named Darral Dean who really laid one on me.

Glasgow was a base located pretty close to nowhere and sometimes we got pretty desperate for entertainment. One night the gentleman mentioned above, Darrel Dean, and I decided to take our wives and see if the local area offered any hope of amusement. Front gate.....big decision, left or right? Since we knew almost nothing of what was to the right, we turned right and headed for Opheim, Montana. Opheim is small and the only establishment with lights on was a bar. We decided to stop in for a few beers. As we enjoyed our refreshments, we struck up a conversation with a couple of cowboys named Red and Shorty who were fresh off the range from branding and rounding up stray cattle. We flyers traded stories with the cowpokes and vice versa.

After consuming quite a number of beers and preparing to depart, I gave the two guys my phone number and an invitation to visit me at my on-base quarters any time they chose. I promised them I would let them tour the base and see inside one of those eight-jet bombers that could carry enough bombs to blow the entire state of Montana off the map. I advised them to call me from the main gate when they wanted to visit. I never expected to see Red and Shorty again. I promptly forgot all about my invitation.

A few weeks later, my wife and I were entertaining a dozen or so Crewdogs with a party one night in our quarters when the phone rang. I answered the phone only to hear a security police sergeant at the main gate say that I had a couple of visitors at the gate who wanted to visit me. The sergeant said I would have to come down and sign the visitors in. I asked who they were, and he stated they were a couple of men from Opheim who said they were Red and Shorty and that I knew them. I almost had a heart attack right then and there! In a panic stricken voice I told the sergeant to tell the men I was not at home and they would have to leave. I explained I was hosting a party for some very important people and could not see them at that time. It wasn't that I didn't want to see Shorty and Red, it was simply...this was not the time.

About 10 minutes later, another voice called back and said I would have to come to the front gate. He said that two men who wanted to visit me had gotten back in their car and had crashed through the gate and headed towards the vicinity of the housing area. He stated at first they would not stop when they were chased by security vehicles with lights flashing and loudspeakers blasting warning messages. Had they not finally stopped, the tires and maybe the men would have been shot, according to the voice. He demanded that I report immediately to the gate to sign some papers before the two men were hauled off to the brig. With that demand, I yelled to my guests that I had to high tail it immediately and head to the gate to rescue two friends who had really screwed up big time! I could see what small potential I might have in the U.S Air Force diminishing by the second.

After exceeding the speed limit considerably I soon arrived at the gate and asked to see the sergeant in charge. The senior sergeant appeared and I asked where Red and Shorty were. The sergeant looked puzzled and said he knew of no ones named Red and Shorty. I did not believe that he knew was he was doing, so I asked the other airmen on duty if they had seen two men who crashed through the gate with a car that went speeding towards the housing area. They all looked aghast at me as if I had lost my marbles, and one asked if I had been drinking! That is when it finally dawned on me that I had been taken for a crazy ride at my own party.

I reluctantly got in my car and, with tail tucked between my legs, headed for home. I dreaded what I would encounter there. Approaching the place I saw everyone in the front yard with several rolling on the grass doubled up in laughter. I really felt about as dumb as one could feel. However, to be fair, I realized I had been the victim of a great joke.

It seems that the story of my friendship with the cowboys had spread to some of my guests. Two of them used a phone next door and disguised their voices to pull off the greatest joke ever played on me. I completely swallowed their joke hook, line, and sinker. My reputation of pulling jokes on others without falling for jokes myself had been tarnished forever!

The author and his crew stand on B-52 " Miss Piggy"

Revenge Is Sweet

Dave Howell

It seemed that some of the most popular movies at any alert facility were those which dealt with revenge. Charles Bronson and the "Death Wish" series were perennial favorites. You know the ones where ole Chuck spent 90 minutes hunting down and knocking off the guys who had killed his wife/daughter/mother/girlfriend. For whatever reason, Crewdogs loved the concept of "getting even." Maybe it was because it was because we were always on the receiving end of every bad deal that came along, and subconsciously we wanted to be on the other end of the stick.

During one icy-cold January alert tour, I witnessed revenge elevated to an art form. No, no one was killed or even maimed. But one junior birdman learned the hard way that it was unwise to publicly speak ill of the enlisted crew force. And absolutely no group was better at getting even than enlisted Crewdogs.

Gunners and boomers were indispensable members of their respective teams, but they were the only crew members who weren't commissioned officers. The gunner protected every life on his bomber, and why even have a tanker if the boom operator wasn't there? Yet, on every crew, gunners and boomers were called upon to perform most of the "menial" tasks. They were expected to pick up the inflight meals, run the projectors, load the coffee urns and gas up the alert trucks. But the worst of it may have been dealing with brown bar lieutenants straight out of training at Castle. The FNG with the little gold rectangle on his collar instantly outranked even the gray-haired master sergeant who had been flying since before the new copilot was potty trained.

However, the enlisted guys were not at a total disadvantage. They were protected by their ability to look out for one another in things like the "Gunners' Union", and they knew the system inside and out. Therefore, life could be hazardous for the brand new officer who demanded respect without first earning it.

I saw this concept in action at Blytheville AFB, Arkansas in the early Eighties. I was the senior A/C on alert when a new bomber copilot, fresh from CCTS, was pulling his first alert tour. This guy was a big, tall, physically impressive dude. His name was Muddy Waters. I knew this because his name tag said "Waters" and the vanity plate on his orange Corvette said "Muddy." He was a rather loud individual, and he certainly "talked the talk" of a sh*t-hot aviator, even if he had not yet proved to his fellow Crewdogs that he could "walk the walk" of the same.

It was the last night of an uneventful alert tour, and all the crews were snug inside our warm mole hole passing the final hours until we could start our three days of freedom. I was "reading" a *"Playboy"* someone had left in the dayroom and Muddy and three other young bomber Crewdogs were playing cards at one table. At another table, a few boomers and gunners were doing the same.

I don't know what card game Muddy and companies were playing, but it obviously required that he and the guy across the table be partners. From the conversation and the laughter, I could tell that Muddy and his partner were on a losing streak. At the end of one hand,

Muddy loudly proclaimed that his partner was "playing like a dumb-ass boomer," which was met with guffaws around his table.

Across the room, at the NCOs' table, no one was laughing. A veteran boomer from south Louisiana stared hard at the laughing lieutenants who were oblivious to the effects of Muddy's insult. I put down my magazine, getting ready to defuse the situation. But, following a head nod from their leader, the enlisted troops filed out of the room without a word.

I approached the lieutenants, told them what had just happened and suggested that they might want to be a little more humble until they had earned some flyboy credibility, but in words a bit more "colorful" than that.

The next morning dawned sunny and bright but bitterly cold. Yes, it even got cold in Arkansas. After breakfast and changeover, we started the bag drag to the parking lot to get a jump on our C-square. As I was throwing my EWO bag in my car, I heard a very loud, very long stream of profanity from none other than Lieutenant Waters. Everyone in the parking lot started to congregate around Muddy and his orange Plastic Fantastic. His 'Vette had mutated into an ugly lump.

His car looked more like an orange(ish) iceberg in a sea of otherwise lightly frosted vehicles. It was covered in ice, and I don't mean a thin sheet of ice. It was frozen to the ground under several inches of solid ice. The ice prevented entry or movement of any kind. No matter how heated Muddy got, his 'Vettesicle was not going anywhere - at least not until the weather warmed up in a day or two. As the crowd figured out what had happened, the scattered chuckles grew into full scale derisive laughter. Someone shouted, "Hey, Muddy, who'd you piss off?" The hulking lieutenant was fuming at being the butt of this very pubic joke.

Recalling the events of the previous night, I was certain I knew who had taken Lieutenant Loudmouth down a peg with such outstanding creativity. I wandered over to the Cajun NCO who had heard Muddy's insult as he was loading his truck, and I asked him, "Sarge, you didn't have anything to do with this, did you?"

His jaw dropped in mock amazement, and with a sly grin he answered, "Who, me? Why, no sir. I'm just a dumb-ass boomer."

My gunner eventually filled me in with the details. Sooner or later, everyone knew the real skinny. The boomers and gunners on alert had organized the "ice down." They had run a hose out to the parking lot, and working in shifts had taken turns spraying the Corvette with a mist that gradually froze the car to the asphalt. It had taken all night, but they had a wrong to right. Revenge is sweet.

The author in 1973.

In the Buff in a BUFF

Zane Walker

The D-Model BUFF was notorious for its quirky air conditioning system. Each plane seemed to have its own "personality" in that regard. The pilots generally could adjust the controls so the crew was reasonably comfortable. There was a control knob at the radar navigator's station intended to allow some extra ventilation in the lower compartment. Sometimes it helped, and sometimes it didn't.

On one particular Arc Light mission with our Glasgow AFB crew of John Durham, the Aircraft Commander and Jackie Thomas, the copilot, we were roasting downstairs. The control knob didn't help, nor did repeated entreaties to the pilots upstairs. I always suspected the pilots would give lip service ("Roger, heat coming up/down, Radar") and kept the pilot's station nice and comfy. After several requests to lower the heat to no avail, I had a goofy notion. I stripped off every stitch of clothing and crept upstairs. It was getting dark, so the pilots didn't notice me as I slipped into the instructor pilot seat behind the pilots - in the BUFF. I tapped John on the shoulder and asked with a tortured expression, "Say, could you cut the heat down a little?" The pilots, after the initial shock, just cracked up.

On another mission with John and Jackie, they had their little joke with the heat - or lack thereof in this case. We were freezing downstairs, and after no amount of polite requests seemed to help. I unstrapped and climbed upstairs. John and Jackie were wearing their heavy jackets and gloves - and having a good laugh at the navigators' expense. Each time we would ask for more heat, they would turn it down a notch. Later, Jackie said what was so funny is that we were being so polite as icicles were nearly forming on the ceiling.

We were highly trained and disciplined crews, but those many long trips across the South Pacific from Guam to Vietnam got awfully tedious. It has been said that "A little nonsense now and then is relished by the wisest men." But after many months of separation from home and family, while enduring the stress of so many long combat missions, laughter was necessary for our mental health. I daresay most Crewdogs relate to the craziness portrayed in the "*M*A*S*H*" TV series.

Get Back JoJo

Harry Tolmich

The wife of a gunner that had died in an aircraft accident at Glasgow AFB wanted her late husband's body shipped to Texas for burial. The dead gunner and his wife owned a travel trailer and she wanted to take it to Texas when she went for the funeral since she would not be returning to Montana.

The squadron commander requested a volunteer to help the widow drive back to Texas and tow this trailer. Sgt Paul Kiviaho was nominated to go since he knew the widow and knew that she would be okay with him on the long trip. She packed up her trailer and they all headed out - her, Paul, and her pet monkey "JoJo".

The monkey was okay on most of the trip, except that he kept messing with Paul as he was driving and trying to pay attention to pulling the travel trailer. JoJo was pulling Paul's hair and pulling his ear and generally making a pest of himself. At a stop after getting fuel, the widow handed Paul money to pay for the gas, I guess JoJo thought they were touching and he bit Paul's hand.

Paul did not make a fuss over the bite since he knew the widow was in mourning. That evening when they stopped for the night, Paul rented a motel room as he had done every night and the widow slept in the trailer with JoJo as usual. In the evening she asked Paul if she could use the motel room's shower and Paul said sure. She left the trailer to shower and Paul was left in the trailer with JoJo. At this moment it occurred to Paul "Hey, I'm alone with the monkey."

JoJo was eying Paul as he got a broom and said "Now you son of a bitch!" He chased that monkey for 15 minutes swinging at him with that broom in that small trailer.

The next day they continued traveling down the highway with the monkey sitting near the rear window of the car. The widow turned to

Paul and said "I wonder what's the matter with JoJo." This is a true story as related to me by Paul Kiviaho.

Sheppard AFB Alert facility in 2007.

A Lot of Horsepower

Gary Henley

I remember watching the old 1949 *"12 O'clock High"* movie while sitting on B-52 satellite alert at Sheppard AFB, Texas. That movie begins with an attorney, the character Harry Stovall (actor Dean Jagger) perusing around an old English antique shop where he stumbles upon an old "Toby jug" from his WWII unit stationed at an air base in Archbury. He purchases the mug and then visits the old base, now barren—a stark contrast between the past (a blur of activity and aircraft noises and the hustle and bustle of bomber Crewdogs, maintenance troops, and staff personnel during the war) and the present moment in the movie. Harry Stovall scans the runways and empty buildings hearing nothing but the wind as it whistles past him. At this point his memory kicks in, and the scene blends the sights and sounds from the present to those days gone by when he was there living through mission briefs and debriefs and sweating bullets as he watched the bombers returning from their hazardous missions over Germany. He then hears the Crewdogs singing their favorite songs as they raise their "Toby jugs" to toast the end of another tough day of flying.

Recently I stood beside our old SAC taxiway next to the Satellite Alert "Christmas Tree" parking ramp off the end of the runway at Sheppard AFB, with my son Stephen (who was going through Undergraduate Pilot Training there), and I relived a surprisingly similar experience. I remember the Satellite Alert stories in the previous *"We Were Crewdogs"* books. I stood there looking out over the now empty ramp as the wind whistled around what used to be our old SAC Alert Facility that is now being used as the Communications Squadron building and for Regional Approach Control (RAPCON) functions. I couldn't help but remember the scene I just described from *"12 O'clock High."* At that moment memories came flooding through my brain. I could hear the noises of B-52 engines, alert trucks and maintenance vans. I could see all those who served there with me including the maintenance troops, security police, and detachment staff members. It was just like in the movie—it was eerie, but somewhat reassuring at the same time. I knew that those memories would never be forgotten, nor would the time I served in SAC sitting alert duty during the Cold War era.

Standing there beside the Sheppard Alert Facility, something flashed into my memory that I had always wondered about. It was one event that I just had to investigate right then and there. That memorable event happened one morning when our B-52D crew had experienced multiple engine problems (so what's new?). I had chosen to stay with the pilot and copilot as they worked with maintenance to get those pesky engine problems fixed on the aircraft. That morning we spent several hours on the ramp while maintenance went through all their system checks and checklists with the pilots to locate and resolve the problems.

Finally, the crew chief gave us the thumbs up to start the engines, which began to spool up one at a time. Since the readings still weren't quite within specs, the pilots were asked to push the throttles higher on the affected engines for about 30 seconds. Our wheels were chocked, and our brakes were locked solid as the power came up higher and higher. The entire aircraft was rocking by this time, and I knew that each of those two J-57 Pratt and Whitney engines were pushing out close to 10,500 pounds of thrust behind us. The noise was deafening. I began to wonder whether or not the brakes and the chocks would hold us in place much longer. After a few seconds, the copilot remarked, "Look at that! Look at that, will you? I've never seen anything like it! That's the funniest thing I've ever seen in my life!" By the time I could get up to the window to see what was going on, all I saw was a huge dust cloud behind the plane and a herd of a dozen wild horses galloping as fast as they could away from our bomber. The copilot said that the horses were actually rolling over and over from the sheer power of the blast of those B-52 engines. Since our engines checked out good at that point, they were then shut down. The crew exited the aircraft and the "no lone zone," area laughing about the rolling horses all the way back to the alert facility.

Now, as I was revisiting Sheppard, I tried to resolve a question in my mind—"How close to our bomber where those horses to be rolled like that from an engine blast? Is it even possible, or was somebody pulling my chain?" Near the alert facility was a restricted area of land which only wild horses were allowed. Beyond this area was a closely guarded road leading to the ammo dump. The wild horses would come up to the barbed wire fence as we were going through our morning alert checklists to satisfy their curiosity about B-52s. On this particular day, about a dozen horses had congregated right behind our bomber. These horses were truly wild; each day on alert, if you started moving toward them, they would turn and run at top speed. The distance behind our bombers to this barbed wire fence from my memory seemed to be extremely close. In addition, there was a road right behind all the bombers and tankers that was used by the security police on their patrols.

On this recent visit to Sheppard, my curiosity was finally satisfied. My memory had not deceived me. I walked over the area and paced off the distance from the back of our plane to the road and then to the fence. It was less than 10 yards from the back of the ramp to the road, and it was only about five yards from the road to the barbed wire fence. The "Snopes.com" website would be proud to know that this B-52D alert "urban legend" was proven true. B-52s can and did "roll" a few horses at Sheppard AFB in the late 70s, and I'm sure that those horses, for the rest of their lives, never came even close to a B-52 with its engines running!

Now, if anyone asks you how much horsepower came out of those Pratt & Whitney J-57 engines on the B-52D, you can tell them, "Enough power to roll over a dozen wild horses in the dirt."

Memories of BUFF Days

Gates Whitaker

A Great Time-saver

Do you remember the Crew Boxes that contained all of our vests and classified materials? They were like an aluminum footlocker that had to be babysat unless signed for and locked in a secure area. It took two people to carry the thing around. Inside, there were six survival vests, one for each crew member. The vests were like fisherman's jackets, with all kinds of gear in pockets and pouches - I think there were seven storage pockets but can't remember what many of the pouches contained, except for the survival radio. And there was a built in holster with a .38 revolver plus two plastic Ziplock packages of cartridges.

When we went up north to the passes or during Linebacker II missions, we wore the vests. I decided to use the smaller bag of cartridges, which contained six bullets, to load the .38. The larger bag contained 25, and I could save time by not counting them all, or always checking that I had not dropped one in the aircraft during the loading or unloading process. I had already lost a wedding ring - probably dropping it while stowing it in my pocket before takeoff. We were not supposed to wear rings while operating aircraft. My wife was not too understanding about my losing the ring and I didn't think SAC would act any better about losing their ammunition.

It was only later when I looked at the bags more closely, that I read that the bag of six were all tracers. I did, from time to time, have better moments than this.

Smokin'

My EW, Ben Askew, got very annoyed with me after a while since I always bummed cigarettes from him. In those days, a lot more people smoked than they do now. I gave it up when I got out, since cigarettes were no longer 20 cents a pack or so, and I no longer had a cushy government job. Well, he decided to get me back in a big way. The A/C was on a health kick and had given it up for a while, and the Nav and Radar didn't smoke, so for a while it was just Ben and me in the front compartment puffing away. The gunner smoked cigars, I think, but nobody knew what he really did back there, except try to figure out how to make Senior and Chief and still be on a crew. From what he said it was not possible, but he was still trying.

Back to the main story: We were flying out of Guam at the time, a 12-hour flight. I took a break after takeoff and climbout and went back to get a smoke from Ben. Imagine my surprise when he gave me a brand new pack of Luckies. Then he gave me a matchbook containing one match and told me that was the last thing I was getting from him that day. And this was from a guy who used my ration card to buy scotch!

Sometimes the Elevator Doesn't Go All The Way to the Top

Always check the hydraulic packs on the D-Model after letting off a passenger, taking off from a taxi-back landing, and playing musical seats in the cockpit. Landings are a lot smoother when the horizontal stabilizer and elevator have more than the backup pumps operating! This was one time that two high ranking IP's were in the right seat and the jump seat said nothing bad to me. They had turned off hydraulic packs 5-10 to save bleed air pressure during the taxi-back. In fact, nobody said anything at all. Ever! We were all just glad that the nose gear is REALLY strong on that model.

Great Meals for the Troops

Getting up in crew quarters before a night mission on Guam, our A/C, Bob Damico was still on his health kick and decided to cook eggs on a hot plate for a good breakfast before going out. The only problem was that the electricity in the building was too weak to heat the eggs. Those eggs looked like the breakfast from the movie *"Rocky I."* This was in the era before we worried about salmonella. Anyway, they had to taste better than the milk on Guam.

We always ordered two water jugs for a mission since a jug of coffee got cold. The special heater cups didn't work and it was pretty bad coffee to begin with. Sometimes they forgot to rinse the old coffee out of the jugs before refilling them with water. Tasty.

We preferred frozen lasagna and manicotti from the base commissary to the flight lunches. The A/C may have been on a health kick but he was still Italian.

Music to Soothe the Savage Dogs

Our Navigator, Jim Spake, was a peaceful , quiet guy. He learned that you could hook a battery powered tape player to the interphone system and we could all listen to music on "private" interphone during returns to Guam on long missions. The batteries would last two to three hours depending on how cold it was downstairs. I can recall flying over the big active volcano in the Philippines, watching it smoke and steam while listening to Pink Floyd. That was a trip.

And Now, Sports Fans

My trailer bunkmate and EW, Ben, and I were going to enjoy the Super Bowl at U-Tapao in 1973. We knew the game was going to be broadcast on Armed Forces Radio Network (AFRN) on a delay in the afternoon. Ben was a southern boy, and a big fan of the undefeated "Miama" Dolphins. Since they were playing the Redskins, and I followed the Cowboys, I was also for the Dolphins. We slept late and ran to get lunch before the broadcast, being very careful not to see or talk to anyone who knew the outcome of the game. Back in the trailer, we turned on the radio about two minutes before kick-off. The signal was not strong and it was not a very good radio, so we wanted to tune it in just right. The first words we heard were "...and three people were arrested after the Super Bowl, won by Miami 14-7." So much for planning. At least the result of the game was good!

Gas Shortage?

In early 1974, after we all returned from overseas deployments, the Air Force needed to save money. Since maintenance costs are directly proportional to hours flown, we had to cut back on flying hours. Even though flying less, nobody wanted to lose fuel allotted to them. If it was not used, it would not be in the budget for the next year. So we flew shorter missions requiring less maintenance- and raised the airbrakes to use up more fuel while flying less. I guess it made sense at the time.

Random Thoughts

In 1,200 hours in the D, F and C – I had one perfect landing and one perfect ILS approach. It gave a new meaning to *E Pluribus Unum*, especially for the other poor crew members.

Is there any wilder looking countryside than Southern Laos?

It is a big surprise when an F-111 "torches" at night near you.

Did anyone ever get the guys in the orange flight suits at U-Tapao to talk about ANYTHING?

Highlight - meeting Bob Hope at U-Tapao. He came in and shook hands with all the crewmembers there right in the briefing room with the missions still up on the map.

Do you remember playing bridge on tanker ferry flights using hand signals to bid?

When hooking up expensive, powerful new speaker systems that have been carried home on a rack in a B-52 bomb bay, allow several days before attempting to operate. They don't do so well if they are still cold-soaked.

Definition of a long day - on leveloff out of Guam, the autopilot will not engage, and it's not a circuit breaker.

Another definition of a long day - a new hot spot in the helmet.

Who stole the last Stratofortress logo from all of the yokes?

Do you remember taking the camo stickers off your helmet and cutting the mission patches off your flight suits at the end of 1973?

Did you think about the kids' story "*The Little Engine That Could*" during a loaded uphill takeoff on the runway at Anderson?

If you screw up real bad, you can still get a job in quality control at Singha.

Beware the Ides of March - My last mission was on a BUY NONE on March 15, 1974 and we passed!

Left to right: 1st Lt Charles K. "Chuck" Geiger, Navigator; Captain Edwin "Sandy" Richardson, Co-Pilot; Captain Larry "Bubba" Guinn, Radar Navigator; 1st Lt Eric "E-dub" Wilson, Electronic Warfare Officer; TSgt "Guns" Schultz, Gunner; Captain Denver D. Robinson, Pilot; and, presenting the award is Lt Col Jack K. Farris, 62nd Bomb Squadron Commander.

Wrinkled and Tucked In

Denver Robinson

I don't know if the designers planned it that way or not, but every BUFF I ever saw had wrinkles on the side of the aircraft.

When I first walked around the bird at Castle AFB, California, where all new crew members go for initial training, one of the first things I did was ask my IP if the wrinkles were okay. He said yes, don't worry about them, they smooth out in flight.

IP's could get away with saying things and doing things that others couldn't say or do. I know because not only had I previously been one, but later in my career I would become a Stan/Eval pilot who gave check/evaluation rides to other crews/pilots. So was a guy I'll just call "Cig" in case someone who reads this cares, and the statute of limitations hasn't run out.

"Cig", like me, had also been a T-38 IP straight out of Pilot Training and then got "sucked" into SAC to fill the voids left because the Air Force had continuously tasked BUFF crews to do 179-day TDYs to the Southeast Asian War Games with VERY short breaks in between. It was kinda like they are doing to troops in the sand bowls today.

Anyway, one fine, clear day in about 1982 he and I flew out to Red Flag as a formation of two B-52s. The third bird died with maintenance problems and was left behind at our home base.

The standard "formation" position for the number two BUFF was one-half mile in trail and 500-foot stacked up (and usually on autopilot). On this day, for some reason, as the pilot of number two, I pushed the mike button and said something like "Two requests clearance to close trail."

For those that might not know, "close trail" in a T-38 was so close it took your breath away the first time the IP demonstrated it. It put your eyeballs about four or five feet lower and around eight or so feet back from the bottom lip of the afterburner nozzles of the lead aircraft. That was just far enough back that if the student pulled too hard you would 'just' fly directly through the jet wash and not hit lead with your pitot tube.

Well, anyway, Cig didn't hesitate two seconds to say "cleared trail" and I moved slightly to the side to avoid his turbulence and moved down and forward-forward-forward into a position similar to that I imagined I'd be in if his aircraft was a tanker. It was not nearly as close as I'd imagine 'real' close trail to have been, but close to close enough to refuel.

After staying there a little while I moved to his right into a "fingertip" formation position where our wings were NOT overlapping, and called "Two's in."

Shortly there came a little wing rock from lead, which was the T-38 (and standard formation) signal to change fingertip position to the other side by performing a "cross-under." This required me to drop behind and below the lead aircraft, cross under, then move up and forward into position on the other side.

I complied and moved to the left fingertip position.

Now, as a matter of background info, normal mission lead change in the BUFF was a complicated procedure requiring the involvement of the radar/nav team and the gunner - who monitored everything on radar. Well, the procedures worked great day or night in the weather or not so therefore were overly involved for clear day, visual formation.

Anyway, "Cig" called for a lead change, I agreed by simply responding "I'm lead!" and "Cig" moved out slightly to the right, then down, aft and back left into fingertip on my right wing. After a while I gave him a wing rock and he moved under and around as I had done earlier.

After a short while we reformed the BUFF style cell and then went on to do our mission.

Oh, I almost forgot why I started this story. The IP didn't know what he was talking about - the wrinkles REMAIN there in flight.

Chapter Six

Lest [lest] - *conjunction* - for fear that.

We [wee] - *pronoun* - oneself and another or others.

Forget [for-get] - *verb* - cease or fail to remember; be unable to recall.

Our Birds of Green

George Thatcher
©2006

It's been forty years, a full two score,
Since we coasted out from the Homeland's shore
On our first foray in time of war,
Which we thought would be just a six-month tour.

We were over-grossed and underpaid,
A bit too cocky to be afraid,
And the ground troops thought we had it made.
They were times to hate, but would never trade.

So we flew our BUFFs from an isle pristine,
And we bombed the foe from heights unseen,
And our thoughts were pure and our conscience clean,
And we ruled the skies in our birds of green.

And we rendezvoused in the dead of night,
With the only birds that could mate in flight,
Embracing, probing with all our might,
Till, full and quivering with delight,
Parted ways as we dipped from sight.

And we often envied those tanker lads,
Who came and silently passed their gas,
Then spent the day with a tall, cool glass,
While we sweated it out over Mu Ghia pass.

We were never supposed to lose a crew,
But war is blind, and more than a few
of the greatest guys we ever knew
paid the price for us; didn't make it through.

So we offer a silent prayer of thanks,
To the crews of the bombers and the tanks,
As we stand tonight, in thinning ranks,
And embrace old comrades on our flanks.

And we'll carry on from our isle pristine,
And we'll bomb the foe in a nightly dream,
With our thoughts still pure and our conscience clean,
As we rule the skies in our birds of green.

Author with Pickle Barrel

The Pickle Barrel Story – A Tribute to Major George Donald Jackson

John Huisjen

On January 1, 1967 the 24th Bomb Squadron, 6th Strategic Bomb Wing of the 16th Air Force met for the last time at the Officers' Club at Walker AFB in Roswell, New Mexico, to celebrate the New Year and to bid farewell. At midnight, the night before, the wing stood down from its nuclear alert mission and the crews went home for the last time. The base was closing and everyone was scheduled to be scattered to the Air Force winds, mostly to other SAC bases and operations everywhere. The two operational bomb squadrons and a tanker squadron were disbanded and the B-52E model aircraft from there were sent to the bone yard for safe keeping and destruction. For me an era of SAC crew duty was closing.

The gathering that day was very festive with good food, drink, speeches, awards, and celebration. One of the awards presented was The Old Pickle Barrel award for Navigation excellence. The trophy was a small pickle barrel with a silver B-52 on the top and included the 6th SAW emblem and a brass plate embossed "Navigation Team Award." Below that was the brass plate which identified Crew S-01 as the recipient. Crew members of S-01 were Donald Wilson, Aircraft Commander; Ronald Larson, Copilot; George D. Jackson, Electronic Warfare Office; John Bynum, Radar Navigator; John Huisjen (yes, me!), Navigator and William Hughes as the Gunner.

Because the bomb wing was breaking up, the pickle barrel was to be ours to take, so it fell to the crew to figure out the final disposition of it. John Bynum, the RN was not there so it was up to George Jackson, EW, and me to see who would have his hands on the award from then on. George recalls that we flipped a coin for it and he won. Because I cried so, we agreed to exchange the award each year by request in a Christmas card with the words "Send me the pickle barrel, COD", those exact words! I wrote the words down on a card as our promise to share the award together.

Neither of us reneged and for 40 years that pickle barrel has passed between us on request, not annually, but whenever the other wanted it. The Christmas cards were annual, always with a report of our lives and those of our families, our whereabouts and our eventual retirements. I built a plywood box for the shipping of the pickle barrel which is well scarred from the abuse it had taken yet it has survived the many trips. Now, it has found a more permanent resting place where it should honor all of those who were a part of that history of the Cold War.

Air Force bomb units created pickle barrel awards from the time of WWII to recognize bombing and navigation excellence. The term "pickle barrel" comes from the claimed capability of the Norden bomb sight to drop a bomb in a pickle barrel from 15,000 feet. The Norden bomb sight brought a high degree of accuracy if operated correctly. The bombardier would locate the desired target through optical sights, placing optical crosshairs on the target, to which the autopilot was coupled, and the bombsight with the crew's coordination would bring the aircraft over the desired dropping point, considering, airspeed, drift, and bomb flight characteristics. Later variations led to radar bomb sights and bombing methods that followed the technology of the original Norden bomb sight.

It was an honor in a bomb squadron to receive recognition for crew navigation and bombing excellence. This Pickle Barrel award received by our crew was for the work of the B-52 Electronic Warfare Officer, his ECM (electronic counter measures) scores, and navigation assistance; the bomb scores of the Radar Navigator, and the navigation scores of the Navigator. For me it was an unexpected honor because I didn't remember such an award until it was presented. George Jackson's take on it was that the trophy sat on the squadron commander's desk and we won it twice in 1966, so it was ours to keep. Crew S-01 really was pretty good at winning those kind of evaluations. We were a standboard-instructor crew and the heart of the crew A/C Don Wilson, Radar Navigator John Bynum and EW George Jackson were the very best. We had also been sent to bomb comp that year so the guys on that crew had a good history in squadron competitions. George Jackson pronounced that crew as the best that he served on so that is good enough for me!

But, as they say, "there is more to the story." When I came in to B-52s in 1964 the crews were made up of aviators from WWII B-17s, B-24s, and B-29s, followed by those from the Fifties periods of B-29s, B-36s, B-47s and B-52s. The whole history of modern aerial bombing was fully intact in those crew members. You could spend hours and hours in conversation about that extended time in history when aerial bombing developed from early WWII into the nuclear deterrence of bombers in the Fifties and Sixties. These men were there from the infancy of bombing to the current state in 1964 of a terrible formidable nuclear foe to any enemy of our country. It was a difficult time in our national history, for which the men who served paid quite a price.

As it often happens if you are really good at something in the service, you may be too good to leave it. Such was the case for most of those B-52 crew members who came into the Air Force in those early days. They frequently came through an aviation cadet commissioning programs and they spent their entire Air Force lives in SAC because once in, they were not shared with the rest of the Air Force. They were "too valuable" to go to Squadron Officers School, Command and Staff School, and other schools of professional development which would lead to timely promotions. Many reached their highest grade of major and served their time exclusively in SAC operations. They often served on nuclear alert for long periods of their lives in remote parts of our country. They were consigned to SAC, the SAC mission, and to the SAC way of life with little hope for any change in their aviation careers.

These are the guys to be honored, those who served so faithfully in SAC during the B-52 nuclear era. I had a small part in it and for a short time was associated with the finest men in existence. Donald Wilson, John Bynum and George Jackson of the crew S-01 were like that, commissioned through aviation cadets and spent their Air Force careers on bomber crews. They are my personal heroes, and they deserve the Pickle Barrel award and recognition that goes with it. George Jackson is my ultimate hero, who was in a B-52 shot down over New Mexico in 1961 by an errant sidewinder missile, launched by an aircraft of the Air National Guard. That he survived was a miracle, which George detailed in a story entitled "Shootdown" in the book "We Were Crewdogs II". Every time that we flew and we landed George would say, "We cheated death again." There are thousands of men like George who believed in their service to their country and willingly served during this uncertain time, now called the Cold War.

In the last 40 years I have seen George Jackson twice – once was in Colorado Springs when his daughter Sally was graduating from the Air Force Academy and was to marry a fellow graduate, Art Crain. The second time was in late May of 2007, when I visited him at his daughter's family home in Virginia. George wasn't doing well physically but was his old self with stories of Air Force life and how he beat me at various wagers, physical contests, crew life, and things that I had long ago forgotten. And of course, he talked about all of his family, children and grandchildren who he dearly loved. We also talked about the eventual disposition of the Pickle Barrel award and decided that it should rest at some appropriate place where it would honor all B-52 crewmembers of our era.

On July 10, 2007, I stopped at the Strategic Air and Space Museum near Omaha, Nebraska, a relatively new facility on I-80. I proposed that the Pickle Barrel would be donated to them if they would appropriately display it and give honor to those who served so faithfully during the B-52 SAC era. Brian York, the curator, said they had heard about pickle barrel awards and would like to display it with B-52 crew items. I left the pickle barrel with them to be a part of their collection.

Two days earlier, on July 8, 2007, George Donald Jackson, Major, retired, United States Air Force, entered his great reward. He fought the good fight and lies at Biloxi National Cemetery with his beloved wife Tina who predeceased him. My Pickle Barrel hero is gone from our world and is now remembered and honored for his service to our country. He, like so many others, was a dedicated B-52 Crewdog to the end.

Memorial Day Speech

Bill Beavers

God gave me the providence to be born into a lower middle class American family. Father was a "blue collar" and Mother was a "stay at home" mother. They cared for me and taught me how to live compassionately with my peers and others. I learned the difference between right and wrong from them and it was enforced by them. They knew where I was, who I was with, and what I was doing. I never lived in a "Day Care" nor had someone else watch over me besides my sister or my mom and dad. If my dad didn't have the money or means to buy something, we didn't have it. Credit cards were unknown. There was respect for elders and others, there was honesty in words and actions and there was love in my family. That was the late 1940's and 1950s in Southern California.

As I grew up I wanted to read and learn all I could about being a USAF fighter pilot - specifically a "night fighter" pilot. Military aviation history, stories of WW II military aviators and RADAR technology were an unquenchable thirst and hunger I had. I wanted to be a part of it by direct participation and by reading and writing part of the history of it. I went through Air Force ROTC in the early 1960s and learned that I did not have 20/20 uncorrected vision. So much for my dreams of being a "night fighter" pilot. By that time anyway, night fighters, per se, had gone the way of the dinosaurs and the loud blasting jets had replaced props. I held the American counter-culture and "hippies" in disdain but was not one to provoke a confrontation with them. The world of military aviation had drastically changed in the way of hardware, but the military profession was still held in honor, the American Flag was respected, and most Americans around me spoke well and felt "good" about our government and our way of life. I wanted to be an integral part of defending those "against all enemies, foreign and domestic."

As one leaves the daily contact and living under the umbrella of a parental home, one senses a need to become connected and belong to some to some other body(ies), group(s), or individuals(s). These associations came for me in marriage and a military operations crew (bombers and missiles for me). My chronological sequence happened military first, then marriage. I found "a home" in the USAF and was fulfilling a childhood dream and hobby in it. I was at the front of the nation's defense, and proud of it.

Then came Vietnam. The scourge of being in a politically fought, politically established, "Rules of Engagement" war helped cause the American public to lose support of its military. Unrealistic enemy body counts and the explosion of drug use among the American military and college age civilians at home did not contribute to raising the image of the American military. In those circumstances, the members of the American military withdrew from American society and enclosed themselves in a protective "shell" for self-defense and self-support. It was in this environment that I found myself serving on a B-52D aircrew in SEA in the early 1970s.

I was proud at last to be a military aviator (sorry Navy and Marine pilots but I use the term here to mean anyone who flies as pilot or aircrew.) I was actively doing what I had been trained to do: navigate and drop bombs. While deployed Arc Light, the crew lived together, slept in the same quarters, ate together, and flew together. A couple of years of doing this made a crew integral and closely interconnected. During flights, you knew what to expect, how to expect, and when to expect aircrew and personal actions by any one of the six members of your crew. You knew their families, personalities, likes, dislikes, traits, and characters - they WERE your family. You also took pride in performing your part of the mission as best you could and doing so as smoothly as you could. You knew the mission and each other's lives, and many on the ground, depended upon you.

A combat wartime environment shared by two or more people simultaneously forms some very tight bonds among both the survivors and their departed buddies. These bonds endure for lifetimes, even if you never see or talk to each other again for decades. "We were there - done that." A very exclusive club was formed around some event and time. You were either there and therefore, automatically a member or you weren't and will always be an "outsider."

I was a member of E-57 as a Navigator and proud of myself and our crew. We flew to North Vietnam in Linebacker I and to Hanoi in Linebacker II. We helped get our Prisoners of War out of North

Vietnam. I would not choose to do it again, but if called upon to do so, I would want to do it with the same guys again.

It was this background that served as the basis for a little talk I was asked to give in Cheney on Memorial Day 2007. My comments were based on my belief in being a member of a fighting unit, and it does not matter if that group is a crew, a squadron, a wing, a command, or a branch of service. We all serve our country together and all depend upon each other's values. Below is the text of that speech.

Cheney, Washington
2007 Memorial Day Speech

I am a Vietnam Veteran. I never set foot on Vietnamese soil but flew a few million miles all over SE Asia dropping bombs from a B-52. I grew up with values of the 1940s and 1950s in the USA and still hold to them today - where my word is my bond, a hand shake is as good as my word, "political correctness" is not my style. We live in "One Nation Under God." The world does not revolve around me or my wants. I earn what I receive (not deserve it without effort on my part), and the "Golden Rule" is a creed I live by.

A few years ago, the VFW had an essay contest for elementary, middle and high school students. The topic was, "Is freedom really free?" Several students here in Cheney wrote essays saying it was free. As we look upon these graves before us today and consider those engaged in wars overseas today, and those held POW, how can we say "Freedom is free." *PAUSE* Think about that. Are you willing to tell those men and women who have given you your freedom, "It is free."? Are you willing to tell the families and loved ones of these vets who have waited at home and had their dreams unfulfilled, "My freedom is free."? Are you so self-centered in today's materialistic American society that you deserve your freedom without earning it?

After the Korean War (that happened "way back" in 1950-1953 for you younger people), the American government drew up a "Code of Conduct" for the American soldier to live by. One of its better known tenants is to give only name, rank, date of birth and serial number. But Article I states:

I am an American fighting man in the forces which guard my country and our way of life. I am prepared to give my life in their defense.

and Article IV, in part, states:

I will keep faith with my fellow prisoners. I will give no information nor take part in any action which might be harmful to my comrades.

while Article VI states:

I will never forget that I am an American fighting for freedom, responsible for my actions, and dedicated to the principles which made my country free. I will trust in my God and in the United States of America.

In these words lies some very BASIC, KEY values of the American soldier. Values I have believed in for my entire lifetime.

1. I am an American fighting man in the forces which guard my country and our way of life. I am prepared to give my life in their defense.

I am prepared to die for my country, our way of life (not some foreign country's way of life) and for your freedom. Can you give me an example of a higher price that I can pay for these?

2. I will keep faith with my fellow prisoners [or soldier's]. I will give no information nor take part in any action which might be harmful to my comrades.

Probably every American soldier, very bottom line, does not expect to die or be wounded; it will be the enemy or the other soldiers with me, but not me. But it is this feeling or attitude that forms a "Band of Brothers" (to use a more colloquial term) or buddy with the deepest commitment (often for a lifetime) that "my life depends upon him and his life depends upon me, I shall not let him down." A soldier may fight for "Old Glory," our flag, but in the heat of combat, he doesn't usually fight for highfalutin democratic ideals that are argued by politicians. He fights to save himself and his buddy's butt! And honor often has a high personal price.

3. I will never forget that I am an American fighting for freedom, responsible for my actions, and dedicated to the principles which made my country free. I will trust in my God and in the United States of America.

Again, Freedom isn't free. "I am responsible for my actions." When you're wrong, you're wrong. Apologize, stand up and take your consequences. Don't hide behind a lawyer that is going to try to convince others that "right is wrong" or "wrong is right," or at a minimum, blur and gray and muddy things so well one doesn't know what is right or wrong. Leave your "me-centeredness" aside and serve the better good of the "Golden Rule" rather than "me."

Wars and battles have been fought for a myriad of reasons throughout time. The eventual outcomes, even when experienced by the victorious, have often not turned out the way they were expected to turn out. Japan and Germany are now our allies; the American South has a cultural flavor all its own while remaining a part of the Union.

I encourage the younger generation to study American history, that our heritage and lessons of the past will not be lost for future generations. The Holocaust did occur. Men have walked on the moon. That these veterans served, were wounded, and/or died for your freedom really happened.

Finally, returning to the American soldier's Code of Conduct, "I will trust in my God and in the United States of America." They did!

At the Wall

Karl Nedela

A plaque at the Arc Light Memorial in Guam states that 75 men lost their lives flying B-52 missions against Communist forces in Vietnam between 18 June 1965 and 15 August 1973.

On the 17th of May the Traveling Vietnam Wall made its way to Killeen, Texas, where I live. The Wall was escorted from Waco by a very large group of motorcycle riders who also accompany funeral processions to the Central Texas Veterans Memorial Cemetery. This copy of The Wall is 3/4 the size of the one in Washington, D. C., and is transported around the country to give everyone a chance to share in the experience. We have been to the one in D. C. three times and it is so impressive and emotional. We have many friends whose names are posted on that Wall.

I was asked to volunteer to be at The Wall to help people locate names of friends and family members on the 18th and 19th of May. I have an Air Force ball cap and a vest with B-52 patches on it and that was what I wore to this function. No one could imagine how many questions about BUFFs that I answered while at The Wall. At noon on the 18th they had special speakers at the ceremony to honor all whose names were on this beautiful black marble. Members of the Order of the Purple Heart were holding flags at this ceremony.

As I walked in, I was greeted by one of the members. He told me that a handshake was not enough to show his gratitude for the BUFFs and for those who flew in them. I received a major bear hug from him and he explained that he would not have been there if not for us. He and his group of 12 soldiers were pinned down in a jungle area by many dozen Viet Cong. They were taking heavy fire and were told that there was so much fire power in that group that helicopters could not enter the area and pull them out. His leader was told to maintain their position and that help would be there soon.

A flight of three B-52s were diverted to that target area and they commenced to clear out the enemy. The bomb drop was very close to this man's group. The "bad guys" were scattered into many pieces and very soon the "good guys" were up and marching back through the jungle to their camp. A couple of his group were wounded, but all made it back safely to friendly territory. I was informed that they would have been a total loss if not for the BUFFs and the work they did that day.

There are more stories about those two days I served at The Wall. A young man, a First Sergeant of one of the many organizations at Ft. Hood, was helping guests to locate names on The Wall. He informed me that he had been a student in my class at Killeen High School many years ago. He now had 21 years in the Army and loved the military. He did not know that I had been in the Air Force and flew as a Gunner in the BUFFs. We had a long talk and the next day his co-worker at the information table told me that this ex-student's father was listed on The Wall. His mother was six months pregnant with him when she received the word that his father had been killed in Vietnam. I wish now that he would have taken me to see his father's name. I said earlier that the Vietnam Wall was an emotional place to visit and this really brought me into reality. As I left my car I filled my vest pockets with Kleenex and each day I had used most of them for the guests and some for myself.

I visited with the gunners whose names were on The Wall - especially my roommate, Charlie Poole. Charlie was the gunner aboard Rose 1, a B-52-D flying out of U-Tapao on December 19, 1972, during the Linebacker II bombing campaign against North Vietnam. At the completion of their bombing run, their bomber was struck by an enemy surface-to-air missile and crashed about six miles southwest of Hanoi. Four of his crew became Prisoners of War, but Charlie and his navigator, Maj Richard W. Cooper, Jr, were reported as Missing in Action. In 1993 and 1994, U.S. investigators of the Joint POW/MIA Accounting Command found photographs, records and artifacts in a Vietnamese military museum that correlated to the crashed B-52. In the fall of 1995, a joint U.S.-Vietnamese team excavated the site where they found B-52 wreckage, crew-related items and personal effects. Exactly 31 years after his plane was shot down in Vietnam, Air Force

Master Sergeant Charlie Poole of Gibsland, Louisiana, was buried at Arlington National Cemetery.

He was a great guy and I remember that he had many hobbies. He enjoyed flying model airplanes and spending time at the lapidary shop polishing stones and semi-precious gems. When his tour was over, his plans were to retire with his family and enjoy them.

This generation of people is not getting any younger and many had to be helped up the ramp to get to The Wall. When we located the names for the people we gave them a pencil and paper so they could make rubbings of the names of their loved one. The hosts had many golf carts to help people to get from the parking lot to The Wall. Another ex-student of mine came with her mother and they were looking for a dear family friend's name. They were crying because of the emotion of the event and that they could not find this person's name. I helped them locate his name. One mother brought her three daughters to The Wall so they could see their grandfather's name there. A man used his cell phone to speak to his dad while standing there. His Dad was stationed at Camp Sami San and had delivered the bombs to be loaded on our planes. He wanted me to talk to his father but his phone went dead so that did not work out for us. One young man was visibly upset because his father wanted to be there but was not well enough to attend. I helped him take some photos so he could show them to Dad and other family members.

One of the highlights for me and many other attendees was the presence of a lovely lady named Dorothy Schafernocker. Her son was killed in action about 40 years ago. Please do not mistake this lady for the one that our press has made a "press heroine". It seems like she is a property owner in Crawford, Texas, next to President Bush's Texas home. Dorothy, or "Momma Nocker" as she is called, is my idea of a "real heroine". Shortly after her son was killed, she started receiving mail from his combat mates. They sent her photos, poems, letters, and all sorts of memorabilia about her son. She was invited to attend many of their reunions and has been at many of the events with The Traveling Vietnam Wall. She only goes if invited, not like the one that asks the press to follow her around. She sets up a tent with photos and letters from her son and his mates and roams around the event with hugs and great encouragement to the ones visiting The Wall. She always greets people with: "Welcome home and thank you for your service". I introduced her to my wife, Liz, and she gave us both a big hug and said: "Wives and mothers deserve much love and thanks also."

"Momma Nocker" was one of the main speakers at the closing ceremonies on Sunday night. Her statements were made to all of us that were in Vietnam. Many people would not visit The Wall because they felt guilty because their names were not on it. Her answer to that was: "Never feel guilt that you are not on the wall. There are 58,256 names on that wall and not a one of them would want your name there. Be at peace with yourself and know that those named there are all being taken care of and are content." What a marvelous lady.

On Sunday night, the 19th, they had the closing ceremony. They had all of the Vietnam Veterans come and stand in front of the Wall while the audience lit their candles. The bagpipes were playing and at the end they played TAPS. The entire group filled the walkway in front of the Wall. It was so impressive and moving.

A few days later my wife was getting some tests at Scott and White Hospital. The young woman who was giving one of the tests saw the opal ring on my wife's finger. Liz explained that I had the ring made for her at Liberty Jewelry Store in Sattahip, Thailand, when I was flying BUFF missions from U-Tapao. The young woman got tears in her eyes and said that she learned so much about the Vietnam conflict while recently visiting the Traveling Vietnam Wall brought to Killeen.

It is amazing how much young people learned from their visits to The Wall. Many were not even born when it was going on and all that is known to them is biased media reports.

A QUOTATION FROM THE DIGNITY MEMORIAL WALL

"*Throughout the Dignity Memorial Vietnam Wall experience, the men and women who we lost in conflict will be remembered. Of the 58,256 names on The Wall approximately 1,200 are still listed as missing in action and 3,415 were from Texas.*"

Lt Col Bornus and crew on his last B-52 flight.

A Father's Stories

Clarence E. Bornus
(As Told To His Son David S. Bornus)

My father, Lt Col Clarence E. Bornus, entered the US Army Air Force in August of 1942. He went into the reserves in October of 1945 and was recalled to active duty in April of 1953. He was a bomber pilot all his career, flying the B-24, B-36, and B-52 (D and H-models). He flew 92 combat missions, 42 in WWII in the Pacific and 50 during a Vietnam Arc Light tour in 1969, as well as many alert flights and missions during the 1950s-1960s. He retired in 1970 with 7,316 flying hours. During his career he was stationed at Rapid City, South Dakota; Amarillo, Texas; Loring, Maine; and Grand Forks, North Dakota. When he retired he was a command pilot with the 46th Bomb Squadron at Grand Forks AFB.

My siblings and I frequently encouraged Dad to write down some of his experiences. He would always say that he was going to "get around to it someday," but first he wanted to check with some of his former crewmates and compare notes, etc. He was always modest about his career. I think he felt that he didn't want to talk about his own accomplishments when there were many others who had done things much more noteworthy. Like many combat vets, he didn't consider himself a hero compared to the ones who never came back.

One day in February of 1993, while sitting at the kitchen table drinking his morning coffee, my dad had a sudden heart attack and was gone. He never did write down any stories. So all we have are a few random anecdotes that he mentioned in passing from time to time, at Air Force reunions, air shows, or in the pharmacy aisle at the grocery store. Here are a few we still remember:

1. My Dad was not a believer in UFO's. The reason, he said, is that "I spent a lot of hours buzzing around up there" and never saw anything that he couldn't eventually explain. He told of one incident when he was on a stateside training exercise involving a group of B-52's flying in formation at night, in a straight line, and it was very important not to break formation. All of a sudden they saw the lights of an "unidentified aircraft" at a high rate of speed, on a collision course. There was no time to think about it and so my dad rolled the plane onto its side to avoid the "unidentified aircraft." Of course this required an explanation of why he broke formation. No one else had seen the "unidentified aircraft," and it did not appear on radar. They got back in formation, saw the "unidentified aircraft" coming at them again, and avoided it again. They thought it was some crazy private plane or something similar, but couldn't figure out why it would "disappear" and then reappear, and didn't show up on radar. Then someone happened to notice that down on the ground there was a highway intersection with a gas station, stoplight, etc. and that the lights from that intersection matched up with the "unidentified aircraft" they were seeing reflected off a cloud ceiling. He said that if they hadn't happened to see that, then it might have ended up as another "UFO" report for Project Blue Book.

2. He mentioned that sometimes tail gunners would have a tendency to sleep in the tail turret. When the B-52 flew to a certain altitude, the procedure required everyone to go on oxygen, and "check in" on the interphone that they were on oxygen. When the tail gunner didn't respond (due to being asleep) the procedure called for an emergency dive to a lower altitude, presuming that the tail gunner had succumbed to oxygen deprivation. He said that the sudden sensation of lying flat on his back usually tended to "revive the tail gunner," who would then quickly check in and report that he was on oxygen...

Speaking of tail gunners, Dad always thought it wasn't right that five of the crew of six were officers and could hang in the Officers' Club and the tail gunner was not an officer so had to go to the NCO club and couldn't be with his crew, being treated as "not worthy" or "below" them when he was just as much a team member as the rest of them.

And speaking of falling asleep, there was an occasion when he was really tired and wanted to take a short snooze like the pilots and copilots often did. It was his turn to snooze so he tried to doze off. After awhile, he woke up, looked over and saw that the copilot was also dozing off! He didn't trust him after that to share the napping.

3. Once he told me that at Grand Forks the "Society for Prevention of Cruelty to Animals" made a stink about the Air Police guarding the B-52's on the flight line with German Shepherds, because the subzero wind chill was so bitterly cold up in North Dakota. So, since it was "cruel and inhuman" for the dogs to be out in those conditions, the luckless Air Police then guarded the planes without the dogs. My Dad said that evidently the SPCA didn't mind the cruelty to the Air Police.

4. At an Air Force reunion at Rapid City, a retired master sergeant praised my Dad's resourcefulness. Evidently my Dad was serving as "officer of the day" or something, and there was a SNAFU when maintenance people screwed up and put out gravel on the runways instead of sand, during the winter.

This was a crisis because there were some planes low on fuel and about to land and they didn't have a lot of time to get the gravel cleaned off the runways. Evidently sand was okay to use since it could be sucked through the B-52 engines without harm, but gravel getting sucked through would ruin the engines. So my dad had a "bright idea" to have some other B-52's taxi over to the runway, rotate around, stand on the brakes and throttle the engines, blowing the gravel off the runways quickly. The maintenance master sergeant who told me this story thought my dad was a genius to "save the day" in such a clever way. In turn, my dad advised that a wise officer would always heed the master sergeants' advice because "they run everything" and always knew what was really going on.

5. My dad was adept at handling in-flight emergencies. We have a copy of an "Officer Effectiveness Report" (AF77) which relates:

"During a recent inflight emergency Captain Bornus accurately assessed the situation and with cool manipulation operated the aircraft on two of the four electrical alternators, at times only having one alternator. Assisting the pilot on the emergency approach to an alternate airfield, Captain Bornus handled the electrical failure making numerous restarts of alternators. He accomplished the landing check list, kept the aircraft cleared visually, made radio calls and used excellent judgment in selection of electrical equipment to be operated. On the final approach the flaps failed and with fluctuating power he advised the pilot correctly against emergency operation of the flaps and computed an alternate best flare speed in split second action. His foresight and constant attention to the aircraft condition and position indicate he is a professional in every sense of the word. His proficiency is far above average for his position." (My dad had piloted a shot-up B-24 to a brakeless landing in WWII, and in civilian life had been an electrician, so his prior experiences may have helped.)

On another occasion, one of my brothers recalls being called out of school and told that our dad was in a dire situation up in the air. They could not get their landing gear to come down. The crew had been trying to solve this problem for some time, burning off fuel, etc. Finally, when fuel ran low, they decided that they would attempt to do a belly landing. My brother recalls that the crew members' families were gathered at the airfield and were listening to the radio calls back and forth from the ground to the aircraft, and watching it approach overhead. As the aircraft was descending to the runway my dad decided to try the landing gear one last time, thinking that perhaps the change in altitude pressure or humidity, etc. might change things, and my brother watched as the landing gear came down and locked into position moments before they landed safely.

6. I accompanied my dad to several reunions around the country. I recall several occasions when the events included a tour of a B-52H and B-1B on static display. Many of the retirees at these reunions were WWII veterans who had not remained in the Air Force afterward. I remember one banquet where we happened to be seated at a table with some of the crew from the B-52H that had been on static display that day, and as the evening wore on the crew, who had spent the day somewhat indulgently answering questions about the B-52 from the "rube" old-timers from the prop-plane days, discovered that my dad had more hours in the B-52 than their whole crew put together. Then they spent the rest of the evening asking HIM questions about the B-52.

7. My Dad once commented that people often were surprised to learn just how much of a bomb load the B-52D could carry fully loaded with the "big belly" modification and wing racks. He said that one had to make significant adjustments as the bombs were dropping, to maintain altitude as the aircraft "lifted" while all that weight fell out. He also mentioned an occasion when to celebrate Christmas the ground crews got permission to paint the bombs, "Merry Xmas," stripes, spirals, etc. in white paint. In our family slides we have pictures of these bombs on the wing racks.

8. He said that in Vietnam the mission planning was pretty hamstrung - their targets would have to go through a long approval process whereby everyone all the way to the "village chief in the jungle" had to give approval for a particular target to be hit, by which time the enemy would have moved their stuff out and be long gone. He said that they often observed “untouchable” enemy trucks and material that were safe over the border in Laos or elsewhere, and he believed that they dropped many bomb loads uselessly "blowing up monkeys" in the jungle.

9. On a wilderness survival training exercise, there was an individual who was very afraid of snakes, and frequently worried about whether there were any around. Someone had found an old wooden cane and had been carrying it. They thought it would be funny to sneak up beside this guy's shelter one night and slowly push the cane along beside him, so that it would sound and feel like a snake crawling in. The guy sat up with a yelp, grabbed the cane, and threw it overhead with all his strength. The cane ended up hanging high in the branches of a tree. Someone must have retrieved it because my dad still had the cane.

10. Many of my dad's missions included long multi-day flights. He had a little 15-pound dumbbell that he took with him, to limber up his arms and legs during a long mission. Some of my earliest childhood memories include my dad sleeping up in the bedroom during the daytime, with an arm over his eyes, after returning from a mission. He used to drink his coffee by pouring the coffee out of the cup into the saucer, blowing on it to cool it off, then slurping it up from the saucer. He always ate fast too. I think it was a habit learned from time spent on alert.

After my dad's death, we found a number of photos, slides, and film strips from his Air Force years, along with his uniforms and medals. They are silent relics hinting at an illustrious past, but we will never know many of the stories behind them. Future generations to whom we hand them down will not be able to appreciate this legacy the way they might have if more information had been recorded.

I have enjoyed the stories in the *"We Were Crewdogs"* series because they give me a taste of the world my dad lived in during those years. I know he would have enjoyed reading those stories himself.

I know everyone who served on a B-52 crew must have some of their own tales. They may think that their stories aren't too interesting, or that others certainly have better stories. But they don't realize that their stories will be irreplaceable gold to their families and subsequent generations when they are gone, and they never know when that might be.

In my dad's pocketbook we found a poem that he had copied, and I think it would be a fitting admonition from him to all of his B-52 colleagues:

The clock of life is wound but once.
And no man has the power
To tell just when the hands will stop
At late or early hour.

Now is the only time you own,
Live, love, toil with a will
Place no faith in 'Tomorrow'
For the clock may then be still.

(He adapted this from the poem by Robert H. Smith)

In honor of those who never returned at all, or those like my dad who are gone without documenting their stories, I hope you all will please "get with it" and tell those stories NOW. Write them on paper, type an email, turn on a tape recorder, use a video camera, or dictate them to someone else. I pray you don't wait another hour. Sit down and do it today! Before your morning coffee!

I still hope that someone who reads there stories here and knew my dad will contact me. I would appreciate you sharing any memories with me, while you still can.

(David S. Bornus wrote this with assistance from: Michael D. Bornus, Susan M. Bornus, Steven G. Bornus, and Doreen F. Bornus.)

Lt Col Clarence E. Bornus

Résumé [rez-oo-mey] - ***noun*** **– a brief written account of personal, educational and professional qualifications and experience.**

The Contributing Authors

From this day to the ending of the world,
But we in it shall be remembered-
We few, we happy few, we band of brothers;
For he to-day that sheds his blood with me
Shall be my brother; be he ne'er so vile,
This day shall gentle his condition;
And gentlemen in England now-a-bed
Shall think themselves accurs'd they were not here,
And hold their manhoods cheap whiles any speaks
That fought with us ....

-Shakespeare's HENRY V

Robert C. Amos

Colonel, USAF (Ret.) A 1958 graduate of the University of Wisconsin with a bachelor of engineering degree in mechanical engineering, Bob entered the Air Force in Feb of 1959, attended Air Force Pilot Training, was awarded his Wings on March 16, 1960. He served for 26 ½ Years flying T-34, T-33, B-52F, F-105D, B-52H, C-47 and T-39 Aircraft. He has 160 Combat Missions including 34 in the B-52F, and 126 missions in the F-105D Thunderchief including100 missions over North Vietnam.

His Air Force engineering and staff assignments included duty as the F/RF-4 Program Element Monitor, Titan III Space Launch Vehicle Avionics Manager, Air Force Studies & Analysis Bomber Branch Chief; Chief, Hq SAC/XOBB, Bomber Tactics Branch; and Director, OJCS/J-34 Nuclear Exercise Branch. He served as Commander, 716th Bomb Squadron, Kinchloe AFB, Michigan; and Director of Operations, 28th Bomb Wing, Ellsworth AFB, South Dakota. In 1979, he received his masters in business administration from Auburn University. Upon his retirement from the Air Force, Bob joined the Boeing Company in 1984, where he is presently the Seattle Phantom Works Deputy Manager of the Strategic Development & Analysis' Studies & Analysis Group.

Bill Beavers

Lt Colonel, USAF (Ret.) graduated ROTC from Fresno State College, California in 1965 with a B.A. in history. He went on active duty to Little Rock AFB, Arkansas in September 1965. He attended basic ICBM school at Sheppard AFB, TX, followed by SAC ICBM crew training at Vandenberg AFB, CA. About three years on Titan II Combat Crew in 308 Strategic Missile Wing/373 Strategic Missile Squadron.

Off to Navigator school at Mather AFB (Sacramento, California) in 1968 (Class 70-05 with [I think] about 60 graduates). Then got married to a young lady I had met at Little Rock (where my initial assignment was as a Titan II ICBM launch officer 1965-68). Then to Navigator-Bombardier school at Mather because I wanted a B-52H ("H," not the older models) model assignment in the southern U.S. (Hell, surprisingly enough, I didn't know there WASN'T any such basing arrangement for the B-52Hs --- they were all in North Dakota,

Northern Michigan --- COLD, brrrrrrr country). Should have gone to ECM school instead with my interest and knowledge about it (my avocation for 50 years has been the development of the American military night fighter airplane and the ground and airborne radars that made them possible, 1940-1955). But that is another story.

Clarence E. Bornus

Lt Colonel, USAF (Ret.) Entered the US Army Air Force in August of 1942. He went into the reserves in Oct. of 1945 and was recalled to active duty in April of 1953. He was a bomber pilot all his career, flying the B-24, B-36, and B-52 (D, H) and other aircraft. He flew 92 combat missions, 42 in WWII in the Pacific and 50 during a Vietnam Arc Light tour in 1969, as well as many alert flights and missions during the 1950's-1960's. He retired in 1970 with 7,316 flying hours. During his career he was stationed at Rapid City, South Dakota; Amarillo, Texas; Loring, Maine; and Grand Forks, North Dakota, with tours of duty in Guam and Thailand. When he retired he was a command pilot with the 46th Bomb Squadron at Grand Forks AFB. After retirement he returned to his family farm in Montevideo, Minnesota, and was active in local and civic affairs. He passed away in 1993.

Walter J. Boyne

Colonel, USAF (Ret.) Former director of the National Air and Space Museum of the Smithsonian Institution, enlisted as a private in the United States Air Force in 1951 and retired in 1974 as a Colonel with more than 5,000 hours in a score of different aircraft, from a Piper Cub to a B-52. For his achievements, in 1998 the National Aeronautic Association named him an Elder Stateman of Aviation.

The Federation Aeronautique Internationale, the international aviation organization of which NAA is a member, honored Boyne with its 1998 Paul Tissandier Diploma.

He has written more than 400 articles on aviation subjects and is one of only a few authors to have had both fiction and nonfiction books on The New York Times bestseller lists. His nonfiction books include The Smithsonian Book of Flight, The Leading Edge, Weapons of Desert Storm, and Boeing B-52: A Documentary History; his fiction books include The Wild Blue (with Steven Thompson), Trophy for Eagles, Eagles at War, and Air Force Eagles. Most recently, he has been instrumental in the development of the Wingspan Air and Space Channel, a 24-hour television network devoted to aviation. Colonel Boyne was inducted into the National Aviation Hall of Fame on July 21, 2007.

James E. Bradley

Lt Colonel, USAF (Ret.) Graduated from college in 1959, with a Bachelor of Science in Mathematics and Chemistry, and enlisted in the United States Air Force in order to avoid being drafted and to preserve a chance of getting into flying training. He entered basic training in July of 1959 at Lackland AFB, San Antonio, TX. He completed basic training in October 1959 and was assigned to the Air Force Special Weapons Center, Physics Division's Pulsed Power Laboratory at Kirtland AFB, NM, arriving in October 1959. There he performed Engineering & Scientific Aide duties consisting of mathematics and computer data reduction aide. This was his first introduction to computers, key punch machines and to scientific laboratory work. The Executive Officer in the Physics Division was Major Lew Allen, later he became General Lew Allen, Chief of Staff United States Air Force. The focus of the work here was finding a way to simulate high-altitude nuclear detonations in a laboratory.

He applied for and was accepted to Officer Training School (OTS) and entered training in April 1961 at Lackland AFB, TX, and after three months of training was commissioned a 2nd Lieutenant on 27 June 1961. From OTS he was assigned to Undergraduate Navigator Training at James Connelly AFB, Waco, Texas. Here he flew on the venerable T-29 Flying Classroom. He received his Navigator's wings in April of 1962.

He was assigned to Electronic Warfare Officer Training at Mather AFB, Sacramento, California, and graduated from EWO schooling in February of 1963. He attended USAF Artic Survival Training at Stead AFB, Reno, Nevada, and entered B-52 Combat Crew Training School at Castle AFB, Merced, California, in March of 1963. All flights were on B-52B/F aircraft. Completed training in July of 1963 and was assigned to Glasgow AFB, Montana, where the B-52D aircraft were assigned.

He remained assigned to the 91st Bomb Wing, 322nd Bomb Squadron as an EWO on a Combat Crew until the unit was disbanded in spring of 1968. Lt Col Bradley has approximately 3,400 hours total flying time with 2,356 hours in the B-52B/C/D/F.

Lt Col Bradley was assigned to the USAF Iron Hand mission known as the Wild Weasel program. He flew in the F-105F as an EWO or Bear assigned to the 355th Tactical Fighter Wing, 333rd Tactical Fighter Squadron stationed at Takhli RTAFB, Thailand. After the Wild Weasel tour he entered the Air Force Institute of Technology's Civilian Institutions Graduate School at Texas A&M University to study Computer Science and graduated in December 1970. He was assigned to Headquarters Air Training Command in the Data Automation Computer Division where he was Computer Operations Branch Chief.

He attended the Armed Forces Staff College and graduated in June 1974. He was then assigned to the U. S. Readiness Command/J5, Joint Operations Planning System Officer, located at MacDill AFB, Tampa, FL, were he was instrumental in the development of the JOPS computer based planning system. From there he was assigned to Headquarters, Tactical Air Command (TAC) on the Planning Staff of the DCS/Plans, BG Larry Welch, later Commander-In-Chief SAC and subsequently Chief of Staff USAF. It is a small world.

After retirement from the USAF in February 1980, he went to work for the Boeing Military Airplane Company, Wichita, KS, on the B-52 Modernization Program. Here he was assigned as a Software Test Engineer in the laboratory where he was responsible for testing the Offensive Avionics System's software in the laboratory prior to flight in the B-52G/H. He also worked on flight simulators and mission planning projects as software test engineer. He retired in 1995. He and his wife, Marian, currently live near Westmoreland, KS, on the farm she was born and raised on.

John R. Cate

MSgt, USAF (Ret.) Was born and raised in Oliver Springs, Tennessee. Attended Clinton Senior High School, graduating in 1971. Attended Tennessee Technological University from 1971 until 1973. With the draft a certainty in my future, I enlisted in the Air Force in April of 1973.

My initial assignment was to RAF Bentwaters, England as an F-4D Phantom II Weapons Release Mechanic. Next assignment was to Lowry AFB, Colorado as an instructor with ATC. After serving as a classroom instructor for five years, I applied for the B-52 gunnery career field. My father had served in the Army Air Corp during World War II as a B-24 Top Turret Gunner/Flight Engineer with the 5th AF and I had always wanted to follow in his footsteps. After graduating from CCTS at Castle AFB, California, I was assigned to the 28th BMS at Robins AFB, Georgia, as a B-52G Gunner. In 1983 I was assigned to the 1st ACCS as an Airborne Radio Operator serving on the E-4B NEACP. I was than allowed to return to the B-52 in 1986 being assigned to the 23rd BMS, Minot AFB, North Dakota as a B-52H Gunner. While at Minot AFB, I attended CFIC at Carswell AFB, Texas in 1987. I served as a B-52H Instructor Gunner and was assigned as S-02 Gunner. I was next assigned to the 43rd BW, Andersen AFB, Guam as the Wing Gunner. I had the honor of serving as the last 43rd BW Wing Gunner. After the deactivation of the 60th BMS of the 43rd BW in 1990, I was assigned as a B-52H Instructor Gunner with the 325th BS, Fairchild AFB, Washington. With the removal of the gunner position from all B-52s in October of 1991, I was assigned to the 964th AWACS, Tinker AFB, Oklahoma as an Airborne Radio Operator, flying the E-3B AWACS aircraft. I served in this position until 1993 when I was medically grounded and removed from flying status. During the last two years of my Air Force career, I served as NCOIC, 552nd ACW Wing Safety Office.

Retired in 1995, I returned to Clinton, Tennessee, where I am employed as an estimator for an established building supply. I have one son who is an architect, living in Dallas, Texas. Serving as a B-52 Gunner was the highlight of my Air Force career and it has given me a lifetime of memories. *C'est La Vie.*

Gerald "Jerry" Channell

Former Captain, USAF I was born in Agawam, Massachusetts, on the banks of the Connecticut River in 1935. The early part of my life was spent following my father around the USA while he was a non-flying instructor in the US Naval Aviation Pre-Flight program. When he was stationed at Pensacola, I used to watch the planes fly around and I think that that is where my love of flying came from. I even got a ride a couple of times. After WW2 ended, we moved back to Agawam for several years and finally moved to Darien, Connecticut where my future wife and I graduated from high school in 1953. I attended Trinity College, in Hartford, Connecticut and was a member of the AFROTC program. I was married to my high school classmate in April, 1956.

Upon graduating and completion of summer camp, I was commissioned a 2nd Lieutenant in the Air Force.

In December of 1957, we (three of us now) departed for San Antonio for pilot training. I completed pilot training in March, 1959, but was held over at Greenville AFB six weeks before starting SAC training. At Lockbourne AFB, Ohio, (four of us now) I flew as a copilot in the B-47, and the crew I was on eventually went into Stand Board. I was certified to give B-47 air refueling evaluations.

We left LAFB (five of us now) for Amarillo AFB in Texas by way of Castle AFB. Didn't do much at AAFB, just pulled alert, checked out as A/C, flew Chrome Domes and training flights. Such was life during the "Cold War" era. Don't remember if I got any medals. Maybe. Probably not!

I resigned my Regular Commission and was granted an Honorable Discharge in December 1966 and shortly afterwards went to work for American Airlines. Raised our three children in Wilton, Connecticut, about 15 miles from Darien where both of us went to high school. In 1983, we moved to Easton, Connecticut about 10 miles east of Wilton.

Most of my flying with American was back and forth to the Caribbean - we had a winter home in St. Croix, US Virgin Islands, for many years. We even experienced three hurricanes!

From the middle 1970's until several years after I retired, I was very involved in youth soccer in Connecticut and Montana as a referee, assessor and administrator.

I retired from American Airlines in April, 1995, and we moved to Red Lodge, Montana. After seven years of endless winter with only a couple of months of "almost" summer and many wildfires, we moved in November, 2002 to the Texas Coastal Bend Town of Rockport.

All in all, my nine-year Air Force tenure was quite uneventful, as were my 29 years with American Airlines. We now have a home on Aransas Bay about halfway between Houston and Brownsville with the Gulf, sunshine and all kinds of wonderful Mexican and sea food to eat. We spend some time each year visiting our three children and six grandchildren in Troy, Michigan; Portland, Oregon; and Parker, Colorado. Retired life is very pleasant!

Doug Cooper

Lt Colonel, USAF (Ret.) Got his commission through OTS in 1965 after graduating from the University of Utah and facing an imminent first draft pick from the U.S. Army. He completed Undergraduate Navigator Training and Electronic Warfare Training at Mather in 1966 and 1967.

His first assignment was to Beale AFB, and the 744th Bomb Squadron. His crew, E-30, was one of the first selected to attend RTU in D-Models at Castle in the summer of 1968 after which he spent six months between the Rock, U-Tapao and Kadena.

After leaving, Beale, he went to Carswell just in time for Bullet Shot and served an additional four TDYs to Guam and Thailand.

After an assignment to the Carswell Command Post (he was the first non-pilot to become a command post controller in SAC), he was reassigned to the SAC Airborne Command Post (Looking Glass). Subsequent assignments at Zaragosa, Spain; Incirlik, Turkey; and Mather AFB were followed by retirement in Sacramento.

Doug and his wife of 39 years, Susan, live in Lincoln, California. His hemorrhoids still ache when he thinks about Giant Lance sorties, of which he had six.

Derek H. "Detch" Detjen

Major, USAF (Ret) Was born in New York City. Grew up there and in the Northern Kentucky/Cincinnati area, entering the Air Force Aviation Cadet program at Harlingen AFB, Texas in late 1960. He achieved the rank of Cadet Major and was the Exec. Officer of Adams Squadron, receiving his 2nd Lieutenant bars and navigator wings in September of 1961. After finishing Electronic Warfare Officer's school at Keesler AFB, Mississippi where he met his wife Betty, it was off to survival training at Stead AFB, Nevada and then B-52 crew training at Castle AFB, California, before his first operational assignment to the D-model at Turner AFB, Georgia. In March of 1966, he participated in the first six-month Arc Light tour to Guam, flying in the first-ever "North" mission on 12 April 1966 with Ellsworth crew E-20, commanded by Lt Col Paul Corn. Returning to Guam in both 1967 and 1968 as a crewmember and staff EWO briefer from Columbus

AFB, Mississippi, he was part of Crew E-13, led by Maj Joe Steele, becoming the first crew to complete 100 Arc Light missions in December of 1967. Crew E-13's 5th Air Medals were presented to them in the briefing room at Guam by the then CINCSAC, General Paul McConnell.

Assigned to Castle AFB, California's Replacement Training Unit (RTU) in July of 1969, he spent probably his most satisfying and rewarding four-year tour, training the G and H-model crews on Arc Light tactics/procedures before their certification and deployment to Guam in the D-model. Four years in the 1st Combat Eval. Group followed at Wilder, Idaho and Ashland, ME, including a five-month stint on Guam at Detachment 24 "Milky," training the recently deployed B-52 wing in ground-directed bombing procedures. His last five years were spent at Barksdale AFB, Louisiana, where he was in charge of B-52 and KC-135 EWO study. A notable event while there was watching the 1980 U.S. vs. Russia Olympic hockey game with the SAC Inspector General team, delaying the onset of their annual visit! He was selected for a final, memorable trip to RAF Fairford, England as part of the 1982 Crested Eagle NATO exercise. His military decorations include a Distinguished Flying Cross, eight Air Medals, a Meritorious Service Medal, two Combat Readiness Medals, an Air Force Commendation Medal and several lesser awards.

After his Air Force days, Major Detjen worked for General Dynamics on the Trident submarine at NSB Kings Bay, Georgia for five years, also attained a graduate degree from Valdosta State University and taught in their on-base education program. A final nine-year stint at Aiken Technical College in Aiken, South Carolina saw him coordinating both the Management and Marketing majors before his retirement in 2000. A lifelong devotee of The Masters golf tournament, he now resides in Evans, Georgia, about 10 minutes from the Tournament, which he has managed to attend for over 40 years.

Bill Dettmer

Lt Colonel, USAF (Ret.) Was commissioned through Air Force ROTC at Rutgers University in 1966. He completed Undergraduate Navigator Training at Mather AFB in 1966-67, and Nav-Bomb Training in 1967-68. His initial operational assignment was as a B-52G navigator at Beale AFB, California.

In 1971 Bill completed Undergraduate Pilot Training and returned as a copilot to the 744th Bomb Squadron at Beale AFB, where he had previously been a navigator. He finished his second B-52 tour as an aircraft/crew commander in 1976 and went into the rated supplement as a Minuteman III ICBM crew commander, flight commander, and subsequently operations officer. In 1980, Bill returned to B-52s as an aircraft commander in the 2nd Bomb Squadron at March AFB, and subsequently Chief of the 22nd Bomb Wing Training Division.

When the 22nd Bomb Wing was converted to an air refueling wing in 1982, Bill was reassigned to Headquarters Air Force Logistics Command at Wright-Patterson AFB, Ohio, directing AFLC's participation in Joint Chiefs of Staff exercises. In 1984, he was assigned to Headquarters 13th Air Force (Clark Air Base, Philippines), where he served as chief of logistics plans and deputy chief of staff for logistics for two years, and chief of military civic actions for one year.

After reassignment to Headquarters 15th Air Force (March AFB) for a year, Bill retired in December of 1988. He subsequently taught graduate courses in systems management for the University of Southern California for seven years. In 1995, Bill founded his own management consulting company, Goal Systems International, which he still heads today. He is the author of six books on various aspects of business management and strategy development.

Richard L. Gaines

Colonel, USAF (Ret.) Retired in 1983 after 27 years of service, all but three of which were in SAC. He piloted B-52D, B-52E, B-52F, B-52G and B-52H aircraft for over 6,000 hours and the F-86 and B-47 for another 1,000. He, his wife Bunny and children Kristi and Lee were stationed at McCoy AFB, Florida; Walker AFB, New Mexico; Grand Forks AFB, North Dakota; Castle AFB, California; Seymour-Johnson AFB, North Carolina; Loring AFB, Maine; and Dyess AFB, Texas. Over those years, Col Gaines held most Squadron and Wing positions including two years as OMS Commander at Seymour-Johnson AFB. He flew more than 50 combat missions as pilot in the B-52D during the Vietnam Conflict. His alert pulling crew duty extended from 1959 to 1976. Among other honors, his crews won the Eisenhart Trophy for the best B-52 crew in 15th Air Force and another was selected to represent Grand Forks in the 1970 Bombing and Navigation Competition at McCoy AFB, Florida. From 1980 to 1983 he was assigned as Director of Operations for the 68Th Bomb Wing at Seymour-Johnson, Director of Operations at the 42nd Bomb Wing at Loring and held the same position at the 12th Air Division at Dyess.

George W. Golding

Colonel, USAF (Ret.) was born in Lake Placid, New York. He graduated from Hudson High School, Hudson, New York and enlisted in the United States Air Force, going to Lackland AFB, Texas, as an Airman Basic. During Basic Training he was accepted into the Aviation Cadet Training Program and transferred to the Pilot Training Class of 61-D attending Preflight Training. He was commissioned a 2nd Lieutenant following Aviation Cadet Primary Flight training at Spence Air Base, Georgia, and Basic Pilot training at Vance AFB, Oklahoma.

His first assignment as an Officer was a B-47E Copilot at McConnell AFB, Kansas, Forbes AFB, Kansas, and Pease AFB, New Hampshire. His next assignment took him to Carswell AFB, Texas, from 1965-68 as a B-52 Copilot and aircraft commander. From 1969-1971 he was assigned to Tachikawa AFB, Japan, and Naha AFB, Okinawa, as a C-13OA pilot and instructor pilot. During this period he flew 287 combat missions. In 1971, the Colonel was a Mission Commander for the 834th Air Division, Tan Son Nhut A B, Republic of Vietnam where he planned and flew as the "On Scene Commander" for 22 Commando Vault missions. Colonel Golding next assignment was to McCoy AFB, Florida, during 1972-1973 where he was a B-52D Instructor Pilot and Quality Control Officer. During this period he flew 115 combat missions over targets in Southeast Asia, including Linebacker I and II Missions over North Vietnam. In 1973 he was the ARC Light Duty Launch Control Officer ("D" CHARLIE) at Andersen Air Force Base, Guam

Reassignment in 1974 took him to the 68th Bomb Wing Seymour Johnson AFB, North Carolina where he started off as the Maintenance Quality Control Officer and then became the Commander of the 68th Field Maintenance Squadron (FMS). He was selected in 1975 to command the 4008th FMS at Barksdale AFB, Louisiana; later becoming the Wing Maintenance Control Officer. In 1977, Colonel Golding was assigned as Director of Logistics Analysis at Headquarters Strategic Air Command, Offutt AFB, Nebraska.

From 1979-1980 he was a student at the Air. War College, Maxwell AFB, Alabama. Upon graduation, he was appointed Chief of Air Warfare Studies and Analysis at the War College. In 1981 he reported to Hill AFB, Utah, to be the Chief of the Resources Management Division, Directorate of Maintenance, Ogden Air Logistics Center.

Colonel Golding became the Vice Commander of the 320th Bombardment Wing in July 1982 and was named the Wing Commander in May 1983. In May 1986 he retired as the Deputy Chief of Staff for Logistics, Headquarters 15th Air Force, March AFB, CA after 27 years of service. He returned to work as a Logistics Analyst and Logistics Manager with the Boeing Military Airplane Company in Wichita, Kansas for four years before retiring with a medical disability. As an active duty spouse, he has been active in working as a volunteer in Wing Protocol, OSC Boards, ALS and NCO school speaker, Family Services and base Chapel activities.

His educational degrees include a Bachelor of Arts Degree from Louisiana Technical University in 1977 and a Master's Degree in Public Administration from Auburn University in 1980. He completed Air Force Squadron Officer School in 1965, Air Command and Staff College in 1973, and Air War College in 1980.

Colonel Golding is a command pilot with over 7,800 hours of flying time. His decorations include the Legion of Merit, Distinguished Flying Cross with V & oak leaf cluster, Bronze Star/ V, Purple Heart, Meritorious Service Medal with two oak leaf clusters, Air Medal with nine oak leaf clusters, Air Force Commendation Medal with three oak leaf clusters, Republic of Vietnam Cross of Gallantry with Palm, and other service awards and decorations.

He is married to Colonel Susan J. Golding, who is the Commander of the 75th Maintenance Group at Hill AFB, Utah.

Gary Henley

Colonel, USAF (Ret.) Was commissioned in 1973 through the Texas A&M University ROTC program. His operational flight experience includes the B-52D, RC-135S (Cobra Ball), and RC-135X (Cobra Eye) aircraft. He served as a flight instructor/evaluator, crew commander, flight commander, operations officer, and squadron commander. He has an extensive background in electronic and information warfare (EW & IW) in the areas of training, systems engineering, and defensive systems flight testing. He served in various staff positions at Wing, Center, MAJCOM, and Agency levels in his career. Unique duties included Assistant Deputy Chief of the Central Security Service; Chief, M04 (National Security Agency); and Chairman of the National Emitter Intelligence Subcommittee (under the National Signals Intelligence Committee).

He earned technical specialty badges in both navigation and intelligence career fields, finishing his AF career as the vice wing commander of the 67th Information Operations Wing at Lackland AFB, Texas, where he retired in 2003 after 30 years of service in the USAF.

Currently, he is the Director for the San Antonio office, Information Science & Engineering Center, of Syracuse Research Corporation and Senior Technical Consultant to the Technical SIGINT Airborne Program Office at the National Security Agency.

Gary and his wife, Becky, reside in San Antonio, Texas, and they have two sons, Stephen and Matthew (both serving in the Air Force) and two grandsons.

Richard E. "Dick" Heitman

Lt Colonel, USAF (Ret.) was born in Lake City, Minnesota in 1934. After high school, attended Winona State Teachers College, Winona, Minnesota for two years. Entered Aviation Cadet program in January of 1954. Received 2nd Lt. commission and navigator wings in 1955 at Harlingen AFB, Texas. KC-97 crew navigator 93rd ARS, Castle AFB, California and 22nd ARS, March AFB, California, 1955 – 1958. Entered pilot training in 1958; received pilot wings at Laredo AFB, Texas in June 1959. B-47 copilot, 22nd Bomb Wing, March AFB, California 1960 – 1963. Transitioned to B-52s, duty stations at Wurtsmith AFB, Michigan and Columbus AFB, Mississippi, 1963 – 1969. Returned to 22nd BW at March AFB, California in 1969 as B-52 instructor pilot. Chief, B-52/KC-135 Stan/Eval Division, 22nd BW, 1971 – 1973. Flew 179 "Arc Light" missions from Guam, Thailand and Okinawa during six TDY tours between 1968 & 1973. Assigned duty as 8th AF B-52D Tac/Eval crew during 1972 – 1973 TDY tours. Logged about 7,000 flying hours during military career; 1,400 hours as navigator and 5,600 hours as pilot (of which 4,430 hours were in the B-52). Retired from the Air Force on 31 Jan 1974.

Gave up my "bachelor career" in 1970 when I married Duchess Rouse. Completed studies for a college degree from CSU San Bernardino, California in 1979. Employed by the City Planning

Department, Riverside, California from 1980 – 1997. Flying activities after AF retirement included 20 years as a volunteer with a local squadron of the Civil Air Patrol as a mission search pilot, flight instructor and check pilot. My interests in flying and things aviation are now satisfied by being a volunteer at the March Field Air Museum, California. We currently live in a retirement community in Banning, California.

Alfred E. "Al" Hodkinson

Colonel, USAF (Ret.) He enlisted in the Army aviation cadet program in July of 1942, completed training and was commissioned in November of 1943 a 2nd Lieutenant, US Army Air Corps and rated as pilot. Sent to B-25 flight training at LaJunta, Colorado and trained in the B-25 as an instructor. Assigned next to Mather Field, California, instructed new pilots until the unit was transferred to Douglas, Arizona. In ferrying a UC-78 as a non-qualified copilot when the aircraft ran out of fuel, they landed on Highway 40 between Tucson and Tombstone. From Douglas he went to Charlotte North Carolina for A-20 RTU and sent overseas to New Guinea. Assigned to the 13th Attack Squadron of the 3rd Attack Group and flew up the chain of islands thru the Phillipines and eventually Okinawa. Flew his last WWII combat mission from Okinawa over Japan the same day they dropped the second atomic bomb on Nagasaki. He could see the smoke but had no idea what is was until he returned to the base,. He became part of the occupation force until returning to the states in December of 1947. He was awarded the Army Commendation Medal for this service.

Promoted to captain in 1946 and became operations officer of the 13th Attack Squadron. Had one trip home during this assignment to ferry an A-26 from Travis to Yokota. At McChord Field he was assigned duties as base operations and special services officer. He attended the Army Assoc Infantry Officer school at Ft Benning, Georgia. Relieved from Active duty in March of 1950 and joined the reserve wing at Portland, Oregon, and flew the C-46 in support of the army troops. He was recalled to active duty April of 1951. His wing went to Ashiya, Japan, in support of the Korean effort. Promoted to Major in 1952 Became an IP in the C-119 and Squadron Ops Officer, 816th Troop Carrier Squadron. Tapped as Adjutant General of the 315th Air Division (Combat Cargo) and moved to Fuchu then Tachikawa, Japan He was awarded the Bronze Star and Air medal.

After a short stint as executive officer of the Air Division, he returned to the states. Became base adjutant at Francis E. Warren AFB in Cheyenne, Wyoming. This was at the time when SAC was installing the ICBMS and the media work was a headache. Went to Fairchild AFB for B-52 assignment and to the 325th Bomb Squadron. He went to Castle AFB for B-52 crew training. He flew as a copilot on a non-combat-ready crew until the pilot tried to land on the main gate at Fairchild. Went with the RO and demanded assignment to another crew and transferred to the 326th Squadron as a standboard copilot. He moved with the squadron to Glasgow AFB and flew as an aircraft commander on many airborne alert missions especially during the Cuban crisis. Promoted to Lieutenant Colonel in April of 1963. He returned to Castle AFB for B-52 IP school. He was then assigned chief, command and control, at Glasgow and remained as a B-52 IP.

With the closing of Glasgow AFB he was transferred to Travis AFB as chief of the command post. He was promoted to Colonel in the reserves in 1964 and returned to Castle for B-52G Difference. With the movement of the 5th Bomb wing to Minot, the B-52s were transferred out and just the tanker squadron remained. He next went to Castle AFB for KC-135 upgrade.. He became the DO of the tanker unit until retirement on August 1,1970. Military aircraft flown with some hours shown: PT-13 Stearman, B-13, AT-6, AT9, P-322 (P-38). B-25 (2,000hrs +) A-20, A-26 (500+), C-47, C-46, C-119, C-54, T-33, T-28, B-52 (2,000+), KC-135 total hours approx 7000. He currently resides at Vacaville, California.

Dave Hofstadter

Colonel, USAF (Ret.) Dave was born in Atlanta, Georgia. He graduated from Mount de Sales Academy in Macon, Georgia, received a BS (Chemistry) from the University of Georgia, an MA (Liberal Arts) from Texas Christian University, and also attended Emory University.

He was commissioned through Officer Training School at Lackland AFB, Texas and completed navigator and electronic warfare training at Mather AFB, California. B-52 training was at the 4017 CCTS at Castle AFB, California. He was then assigned to the 9th Bomb Squadron, 7th Bomb Wing, Carswell AFB, Texas. From there he served four Bullet Shot tours as a B-52 EWO at Guam, Thailand and Okinawa, flying 118 combat missions and accumulating 506 days in theater. Next, he was assigned as an instructor in the 4018 CCTS also at Carswell, training new BUFF EWs. He next was selected to be aide-de-camp to the 19th Air Division commander at Carswell.

SAC Headquarters, Offutt AFB, NE was the next destination. Dave was an acquisition officer for B-52 avionics and then wrote requirements for the planned B-1B bomber in development.

Dave next was assigned to the Pentagon, Washington, DC as the B-1B program element monitor (PEM), serving first in the Air Staff then in the Air Force Secretariat. He was the Pentagon point of contact during the delivery of the entire B-1B fleet. From there he went to the Defense Systems Management College in Ft. Belvoir, Virginia.

Wright-Patterson AFB, Ohio, was next, as director of projects for the new B-2 bomber, serving in the B-2 System Program Office (SPO) in Air Force Systems Command. At the end of this tour, President Bush presented the B-2 SPO with the Collier Trophy, now on permanent display at the Smithsonian National Air and Space Museum in Washington, DC.

From there he went to Air War College at Maxwell AFB, Alabama, an assignment which included a short tour in Moscow, Russia.

The next station was Tinker AFB, Oklahoma, where Dave was the B-2 System Support Manager, Oklahoma City Air Logistics Center, acquiring the support equipment, tech orders and spares and laying in depot planning for the new still-classified bomber. He presided over the construction and operation of the two largest buildings in Oklahoma with no windows. At the end of Dave's tour here, the Air Force presented the B-2 team the USAF Schriever Trophy.

Final assignment was to Electronic Systems Center, Hanscom AFB, Massachusetts as Air Force Product Group Manager for Information Warfare. The job there was to turn the tables on international hackers.

Dave's decorations include the Distinguished Flying Cross, Legion of Merit, Republic of Vietnam Cross of Gallantry, with Palm and eight Air Medals. He is Department of Defense Level III Certified in Acquisition Logistics and in Program Management.

While at Mather, Dave married the former Diane Tomasini of Sacramento, California. At Carswell she presented them with their daughter April, who is now married to Josh Shiflett and they live in Dallas, Texas. While at Offutt, Diane gave birth to their son Dan, who lives with his wife Jill in Chicago, Illinois. Dave returned to Tinker to retire. He and Diane live and work in Oklahoma City, Oklahoma.

Dave Howell

Major USAF (Ret.) Was born and raised in Memphis, Tennessee. Upon graduation from Memphis State University in 1968, I was commissioned and sent to Sheppard AFB as an Administrative Officer. It didn't take long to figure out that flyboys ran the Air Force, so after a few years, I finally landed a pilot training slot (no mean feat since I wore glasses and was pressing the age limit.)

After UPT, I spent about 13 years in SAC, primarily at Barksdale and Blytheville. My progression and assignments were typical of SAC at the time. If all my alert tours were placed end to end, my time behind the barbed wire would total well over three years. I started as Copilot Y on Derrick Curtis' crew in the 596th Bomb Squadron, and left SAC as Chief of Stan/Eval at the 97th Bomb Wing. I did one last vacation tour at USAFE HQ, War Plans, as the resident "Bomber Guy."

Fresh out of the Air Force, I had an incredible stroke of good luck by getting hired by American Airlines (and continuing to fly antique Boeings.) I eventually had to upgrade to modern aircraft, but hey, the flying was fun and there was no punishment. I have now retired twice, and I love it. I have also been blessed with a real trooper of a wife and two great kids.

John H. Huisjen

Colonel, USAFR (Ret.) In August, 1962 I entered USAF Officer Training School at Lackland AFB, and was commissioned in November of 1962. Assigned to Undergraduate Navigator Training at James Connally AFB, then in November of 1963 to Undergraduate Navigator Training at Mather AFB. In the summer of 1964 to survival training at Stead AFB, academics only at Castle AFB, and to my first SAC duty station, Walker AFB, Roswell, New Mexico, in November of 1964.

On January 7 1965, I was assigned to my first crew, made up of some really interesting characters, an exceptionally fine crew, some of whom I have written about. I was moved to another crew for a time and then back to the first crew when it was selected for Bomb-Comp training and competition in the fall of 1966. During that time we had an exceptionally heavy flying schedule and I developed a back injury from which I did not fully recover and in January 1967 I was permanently grounded from flying duty.

SAC still had a permanent hold on me and when Walker AFB closed in 1967 I was assigned to Grand Forks AFB and served there as a personnel officer until I separated from active duty in August of 1969 to begin law school at the University of Denver. While in law school I found a reserve assignment available to me at the Air Reserve Personnel Center at Denver and served there specializing in individual ready reserve management. When I reached 60 I retired from the Reserves.

While in the Air Force I spent at least three of my seven years in various schools for commissioning, flying training, up-grading, professional schools and personnel schools. I am still impressed with the quality of those Air Force schools. For me they were much more difficult and demanding than my later education in law and I attribute the success of my 25 years as a practicing lawyer to my Air Force years of education

I am now retired, living in Fort Collins Colorado, and spend winters at Green Valley Arizona. I am reminded many times of my Air Force days and the fine people in SAC who were "SACumcised" whether they liked it or not, and who served our country so faithfully. I am honored to have served in that era and feel compelled to contribute my recollections of my heroes of the cold war.

John W. "Bill" Jackson

Colonel, USAF (Ret.) He enlisted in the aviation cadet program in August of 1942. He was in the class of 43-J and was awarded pilot wings and a commission as a 2nd Lieutenant US Army Air Corps, November of 1943. At that time as an army pilot he was told he could fly any type of plane that was available! Sure. Upon completion of B-17 phase training as a copilot, he was assigned to the 95th Bomb Group, (first B-17s over Berlin), 8th Air Force. He completed 35 missions July 1944. He participated in the air assault on D-day and the first 8th Air Force shuttle mission to the Soviet Union. He was awarded the DFC, Air Medal with 4 clusters, and the ETO medal with four battle stars.

After VJ day, he was transferred to MacDill Field, Tampa, Florida. His last assigned duty at MacDill AFB was that of a flight supply officer. His account included the base cannon. He was released from active duty in January of 1947 with the rank of Captain. He was active in the Air Force Reserve and completed four summer active duty tours and was recalled to active duty in May of 1951. He was then assigned to SAC, 43rd Bomb Wing, Davis Monthan AFB, Tucson Arizona. He served as a B-50 copilot for two years and then upgraded to aircraft commander in July of 1953.

When the B-50's were sent to the adjoining bone yard, he was transferred to the other squadron on base in the 303rd Bomb Wing, just getting combat ready in their new B-47's. He completed B-47 transitions school at Pine Castle, Florida, in March of 1954. His crew was selected for upgrading to the B-52 program in October of 1957. Upon completion of CCTS at Castle AFB, California, he was transferred to the 325th Bomb Squadron, 92nd Bomb Wing, Fairchild AFB, Washington. He participated in the seven-month airborne alert test, flying 22 missions.

In January 1961, he was promoted to Major, and promptly was transferred to the 326th Bomb Squadron. When SAC renamed its units for World War II units the Glasgow unit became the 91st Bomb Wing, 322nd Bomb Squadron.

He upgraded to IP in 1963, served in the standardization section and was one of two aircraft commanders sent to Guam for indoctrination in Arc Light procedures prior to the wing's scheduled six-month tour beginning in September 1966. He was promoted to Lieutenant Colonel in the spring of 1966. He flew several Arc Light missions before becoming chief of the crew scheduling section.

Upon return to Glasgow, he was assigned as chief of the programs and scheduling section. He was transferred to Beale AFB, California, in February of 1968, and after attending B-52G difference upgrading at Castle AFB, he was assigned to the scheduling section while still being a member of the bomb squadron.

He was promoted to Colonel in the reserve forces in the spring of 1969. Choosing to remain on active duty as a Lt Col he transferred to the B-52 operation in Thailand in November of 1969. However, the President decreed that all reserve officers with 20 or more years of service should be retired in six months. He retired on 31 March 1970 having logged more than 7,100 hours of flying time with 4,100 in the B-52B, D and G models.

He completed work towards an MS degree in Gerontology after he retired, receive his BA degree in Social Welfare and Corrections, and was over half way to an MA degree from Chico State when he decided to call it quits. He currently resides in Yuba City, California.

George Donald Jackson

Major, USAF (Ret.) Resided in Spotsylvania, Virginia, until his death on July 8, 2007. He retired in Richmond, Kentucky from the US Air Force in 1976 after a 21-year career. His post-retirement activities included teaching, real estate, auctioneering, and golf.

George grew up in Richwood, West Virginia (a small mining town back in the mountains). He was the oldest of four kids with two brothers and one sister. He graduated from Richwood High School in 1952 and went to work in the coal mines doing various jobs including surveying. During a strike at the mine, George took an Air Force Aviation Cadet entrance test and entered Aviation Cadets in 1955.

Due to eye vision issues, George was switched from pilot training to navigator training, graduating in December, 1957. He then attended Electronic Countermeasures (ECM) School at Keesler AFB, Mississippi. Upon graduation he was assigned to 75th Bomb Squadron at Loring AFB, Maine. After two years in Maine, the wing broke up, and the 75th was re-assigned to Biggs AFB, Texas in 1959. In 1963, George was assigned to Walker AFB, New Mexico. In 1967, he was assigned to Westover AFB, Massachusetts and then to Richmond Radar Bomb Scoring (RBS) Site in 1973. George served in various leadership positions including two tours in Standardization and Evaluation (STAN BOARD), Electronic Warfare Instructor, and RBS Operations Officer and Commander. Major Jackson accomplished five deployments to South East Asia from 1967 through 1972.

After retirement he attended Eastern Kentucky University and earned a BS in Business Education. (With High Distinction—he adds!)

George met the love of his life, Tina Marie LaRosa while in Biloxi, Mississippi in 1957. They were married in January 1959 and raised three children who were all born in El Paso, Texas. The oldest, Diana now resides in Lexington, Kentucky; Sally lives in Spotsylvania, Virginia, and Donny lives in Atlanta, Georgia. George enjoyed all eight of his grandkids. George wanted to dedicate all his stories to his late wife, Tina Marie LaRosa who stood by him all those years. She passed away July 22, 2004 after a long illness. George passed away on July 7, 2007, and was buried in Biloxi, Mississippi.

Lothar "Nick" Maier

Major, USAF (Ret.) was born in Buffalo, NY. He entered pilot training with Aviation Cadet Class 55-M, in January of 1954, and was commissioned at Williams AFB, Arizona, April, 1955. Immediately after graduation, he was one of the first 2nd Lieutenants to enter SAC's Pilot AOB (Aircraft-Observer-Bombardier) course at James Connally AFB, Texas, and received a Navigator rating. Assigned to B-47s at Smoky Hill AFB, Kansas, where in 1956 his crew was the first from the 40th Bomb Wing to be assigned to B-52 upgrade training at Castle AFB, and subsequently remained there in the 93rd Bomb Wing training cadre.

Nick was a B-52 aircraft commander for 20 years, flying the B through G model aircraft. He received a SAC Crew of the Month Award for an aircraft save in 1967. Served one B-52 Arc Light tour in 1969 with 70 combat missions, and was 8th AF Senior Controller at Andersen AFB, Guam, during Linebacker II in 1972. Retiring as a Major in 1977, he worked 16 years in Travel Industry Management. He is married to Mary Beth, and their son Robert has an Instrument Rated Commercial Pilot's license.

John Mize

Major, USAF (Ret.) After a couple of years as an Admin Officer, I was assigned to be an OTS Recruiter in the upper Midwest. (This was just before Vietnam started to heat up and the warmth of Texas was appealing to many in the midst of a Minnesota winter.) After about a year of that, I began to believe my own story and applied for Pilot Training. Having hung over the fence at Barksdale as a kid, of course, my first choice was the BUFF.

Assigned to Ellsworth in January of 1968, I soon began the endless TDY routine and experienced those many moments of boredom, punctuated by sheer terror that we all went through. I ended up flying 295 Arc Light missions, the final one being the 27th of December 1972 during Linebacker II. Since I had accumulated enough "fruit salad" to wear, that seemed to be a good time to go home!

I spent the next 12 years in various flying and maintenance positions and retired in 1984. I was able to put some of my experience to use when I worked as a Console Operator for the B-52 Flight Simulator - kind of like a life size video game! I retired for the last time in 2003 and, with my wife Joan, am enjoying retirement in Belmond, Iowa.

Arthur Craig Mizner

Major, USAF (Ret.) is a B-52 Combat Veteran Command Instructor Pilot of Vietnam (1969 - 1973) Arc Light, Linebacker I and II missions, and a B-47/B-52 Veteran Pilot of the Cold War ground and airborne nuclear alerts. Craig entered USAF active duty in 1954 as an Aviation Cadet (class 56Q) and was commissioned a 2nd Lieutenant and Jet Fighter Pilot on 28 June 1956 and retired 1 January 1977 with 264 TAC and SAC combat missions, 9206.7 hours in the B-52 and over 11,000 military flight hours. In addition, Craig has over 2,000 hours as a commercial pilot. Craig joined General Dynamics/Lockheed Martin Aeronautics in 1979 and was the F-16 Pilot Flight Manual Manager for the first 17 years. Since than, Craig using his computer skills works as an Advance Technology System Integrator Aerospace Staff Avionic Engineer in support of the F-16 Fighting Falcon. Craig has BS in Industrial Technology from Texas A&M – Commerce, Texas December of 1978 and an AA in Social Science from Tarrant Country College, Texas May of 1998. Craig was inducted in the Phi Theta Kappa International Honor Society on 20 April 1998. For Craig's involvement in the bombing of Hanoi, NVN for the return of all POWs, read "The Eleven Days of Christmas: America's Last Vietnam Battle" by Marshall L. Michel. In May 2002, Craig and his copilot Donald Allen Craig were interviewed and video taped for a TV feature on the bombing of Hanoi. The TV feature should air in the near future. See URL for details: http://www.teleproductiongroup.com/12_72-main.html be sure to navigate to all areas. "Those Who Lived it." The first picture you will see is below. Look at the fifth interview. Craig is a contributing author to the November 2006 Smithsonian Air & Space magazine article Cuban Missile Crisis 27 days at DEFCON 2. In addition, Craig has authored many articles for "*We Were Crewdogs*", a volume of books dedicated to The B-52 Collection as edited by Tommy Towery. In August 2007, Craig did a three hour video interview with the National Geographic Channel Towers Productions for a three hour video about the Vietnam War. The TV showing is scheduled for December 2007. Craig is still working full time and now has over 28 years with Lockheed Martin Aeronautics located in Fort Worth, Texas.

Dwight Moore

Lt Colonel, USAF (Ret.) Is a native of Kansas City, Kansas. He received his B.S. in Education (cum laude) from Kansas State College of Pittsburg in 1968. He entered the Air Force through Officer Training School at Lackland AFB, Texas, in 1968 following a brief summer spent as a bartender in the Imperial Hotel, Cripple Creek, Colorado but only after receiving a draft board notice to report for an Army physical. He completed pilot training at Reese AFB, Texas, trained in the B52F at Castle AFB, and was assigned to the 20th BS, 7th Bomb Wing, Carswell AFB, Texas in the B-52C&D aircraft. While there he participated in Arc Light, Linebacker I and II operations, flying 216 combat missions over Vietnam and Southeast Asia – almost all while on crew Carswell E-15 first as copilot and later as aircraft commander. Six of those missions were flown as an aircraft commander in Linebacker II. Following duties as a simulator instructor pilot (B52D&G railroad car), an instructor pilot, and a flight examiner/Stan Eval with countless weeks of nuclear alert at Carswell in the B-52D, he entered the University of Kansas law school in the Air Force Funded Legal Education Program, receiving his Juris Doctorate degree in 1977.

As a judge advocate he was assigned to Nellis AFB, Nevada; HQ 13th Air Force at Clark AB Republic of the Philippines; back to Carswell AFB, Texas as the base Staff Judge Advocate where he watched the last B52Ds retire; HQ AFSPACECOM, Peterson Field Colorado; and finally as Chief, Contract and Air Law, HQ Military Airlift Command/Air Mobility Command at Scott AFB, Illinois. While at Scott AFB he was the senior contracts attorney for the MAC and participated in the first ever activation of the Civil Reserve Air Fleet (CRAF) for Operation DESERT SHIELD/DESERT STORM. He oversaw the resolution of millions of dollars in claims by the air carriers following DS/DS, and was house counsel for the USAF in the Pan American Airlines bankruptcy recovering $108M for the US

Government (28% of claim but unfortunately didn't get a percentage cut for his efforts).

He retired after 25 years of active duty service in the Air Force in September of 1993. Decorations include the Distinguished Flying Cross with two Oak Leaf Clusters, the Defense Meritorious Service Medal, the Meritorious Service Medal with three Oak Leaf Clusters, and the Air Medal with twelve Oak Leaf Clusters.

He was then hired into DOD Civil Service and began his duties as Chief, Fiscal and Civil Law Division, for the USTRANSCOM Office of Chief Counsel in January 1994. He was promoted to GS-15 in November 1995. He is responsible for civil and fiscal law issues for a combatant command and advises on government transportation and insurance issues peculiar to commercial transportation, international agreements, writes legislation, and administers the command's ethics and environmental law programs. As the DOD insurance point of contact for the Civil Reserve Air Fleet program, he works with the US air carriers, NATO, the Department of Transportation and the Federal Aviation and Maritime Administrations to make improvements to both airlift and sealift insurance programs. He participated in the second successful activation of the CRAF program for Operation Iraqi Freedom/Operation Enduring Freedom and is the command advisor on DOD commercial aviation safety issues.

31 of the legislative proposals he drafted and defended through the legislative process were approved by Congress and enacted into law (he also holds the DOD record for most rejected proposals - 66), mostly aimed at transportation issues, but including legislative changes to give the Frequent Flyer miles earned on official travel to military members and employees (2002) and the authority for DOD to contract for full replacement value protection on military members and civilian employees Household Goods shipments (2003) which will be fully implemented by March 2008.

He is quoted in *"The Eleven Days of Christmas, America's Last Vietnam Battle,|"* by Marshall L. Michel III, Encounter Books, 2002; was a photo contributor to *"Boeing B-52: A Documentary History",* by Walter J. Boyne, Jane's Publishing Co, 1981/Schiffer Publishing Ltd., 1994; and had his photo appear in the October 1996 issue of Air Force Magazine, Vietnam Scrapbook. He is a proud member of the Red River Valley Fighter Pilot's Association (River Rat).

He has been happily married to his college sweetheart, Nona, for 39 years. They have two children and two grandchildren.

Karl D. "Ned" Nedela

Senior Master Sergeant, USAF (Ret.) Was born and raised in Crete, Nebraska. After high school, attended Doane College in Crete and played two years of football. In 1951, I knew I was about to be drafted (not by the NFL) so I enlisted in the Air Force.

After a year in Electronics Courses at Lowry AFB, I attended B-36 Gunnery School in Denver. I went through B-36 Transition at Carswell AFB. Our group was sent to Loring AFB and I served as an instructor gunner for 3 years. In 1956, I was sent to B-52 Gunnery School at Lowry AFB and then to Castle AFB. Returned to Loring and our Squadron was sent to Biggs AFB, El Paso, in the Summer of 1959.

Reassigned to Dyess AFB, Abilene, Texas in 1966. In June of 1970, I went for seven months to U-Tapao as a TDY Wing Gunner.

I graduated from McMurry College in 1972, thanks to the Bootstrap Program. Graduated in August and in September went as a crew gunner to U-TAPAO. I returned in April of 1973 and retired in July of 1973. We moved to Killeen, Texas where I taught and coached for 23 years.

I have been married to my wonderful wife, Elizabeth, for 53 years and have two children and three grandchildren. I have fond memories of my Air Force career.

Jeffrey J. Parker

Colonel, USAF (Ret.) Graduated from Warroad High School, Warroad, Minnesota. He earned his BA degree at the University of North Dakota, and an MS degree from Embry Riddle Aeronautical University.

He received his wings at Laughlin AFB and then attended B-52 training at Castle AFB. His first operational assignment was with the 744th Bomb Squadron, Beale AFB, and he soon deployed to Andersen AFB, where he served as a B-52G coilot with the 64th Bomb Squadron during Arc Light/Bullet Shot. In January of 1976, he was assigned to the 34th Bomb Squadron at Beale AFB, where he upgraded and pulled the last B-52 alert tour at Beale. In August of 1976, he headed north to the 644th Bomb Squadron/410 Bomb Wing, K.I. Sawyer AFB, where he performed as a B-52H aircraft commander, instructor pilot, and standardization evaluation pilot. While there, he graduated from the USAF Instrument Pilot Instructor School.

In 1981, he headed to Castle to be a CCTS IP and was later an instructor in the SAC Instrument Flight Course. He moved to Offutt AFB in 1984 for his first taste of a staff job. Concurrent with his duties at Offutt AFB, he was selected to perform an additional duty as USAF Presidential Advance Agent, a job that took him to a variety of worldwide locations in support of the President and Air Force One.

Returning to Castle in September of 1988, Col Parker became the first operations officer of the newly formed 330th Combat Flight Instructor Squadron (old CFIC), and in March of 1990, became its second commander. In November of 1990, he assumed command of the 328th Bomb Squadron and in June of 1991, became Deputy Commander, 93d Operations Group, with concurrent duty as the initial commander of the newly organized 93d Operations Support Squadron. In March of 1993, he transferred to Headquarters 12th Air Force, Davis-Monthan AFB, where he served as Commander, 612th Combat Plans Squadron until November of 1994.

From November, 1994 to July, 1997, Col Parker was Vice Commander, 5th Bomb Wing, Minot AFB. In Jul 1997, he headed to a better golfing climate and was Silver Team Chief, Office of the Inspector General, Headquarters Air Combat Command, Langley AFB, where he became "one of those" guys. Colonel Parker is a command pilot with more than 4,000 flying hours in the B-52. He currently is the Roseau County, Minnesota Veterans Service Officer and flies part time for Marvin Windows and Doors. He lives in his hometown, Warroad, Minnesota, and invites any former Buff guys to come up and go fishing with him.

Harold E. Pfeifer

Major, USAF (Ret.) Born Nov. 13, 1929 in Urbana, Illinois. Attended University of Illinois, BSME, 1947-1951. Entered USAF (2nd Lieutenant ROTC) in August of 1951. Completed Navigator training at Ellington and Mather in 1952. Completed B-26 crew training (Langley) and flew combat in Korea in 1953. Returned to Altus AFB,

Oklahoma as a KC-97Navigator in 1954. Navigator upgrade and reassignment to Carswell AFB, Texas, 1955. B-36 crew member as Co-Observer, Navigator, and Radar Operator, 1956-1958. Two B-36 aircrews (26BS & 98th BS/11th BW) won SAC Nav-Bomb Competition, 1956. Was the stargazer (co-obs) on the 26BS crew.

B-52 Special training at Mather AFB and B-52 training as Radar Navigator (RN), Castle AFB, California, 1957. RN at Altus AFB, Oklahoma and Wright-Patterson AFB, Ohio, 1958-1962. Attended the University of Oklahoma, Masters in Mech Engr (1962-1964). Project Officer with munitions development, Air Force Armament Test Lab, Eglin AFB, Florida (AFATL, 1964-1966). Assigned to Lawrence Livermore National Laboratory (LLNL), Livermore, California for specialized training, 1966-1968. Returned to AFATL and served as Project Officer and Branch Chief in munitions development (1968-1972). Remained on flying status for 21 years. Retired at Eglin AFB in August of 1972. Returned to full time work at LLNL in 1972. Completed a second career with LLNL and retired in November of 1993. Still resides in Livermore, California. Married, with two grown children, and two grandchildren.

Denver D. Robinson

Lt Colonel, USAF (Ret.) Was born in Memphis, Tennessee only four days into "the baby boom" on January 4, 1946. He graduated from Harding Academy of Memphis and then from Memphis State University with a Bachelor of Science degree in drafting and design from the Industrial Technology Department with a minor in art. He subsequently earned a Masters of Public Administration from Auburn University in Montgomery. He completed Squadron Officer School in 1975 and Air Command and Staff College (ACSC) in 1983, both in residence.

Commissioned through the Reserve Officer Training Corps program at Memphis State University in January 1969 he reported just days later to Laughlin AFB, Texas for undergraduate pilot training. Upon completion in April of 1970 he attended instructor pilot training at the Air Training Command's T-38 instructor pilot school at Tyndall AFB, Florida. He then returned to Laughlin where he was a both an in flight and classroom instructor. Later he rose to be a "check pilot" responsible for intermediate and final in-flight qualification evaluations of student pilots in the final phase of their training.

In 1973 he was reassigned to the Strategic Air Command (SAC) and reported to B-52 pilot qualification training at Castle AFB, California, where he was trained in the B-52 F-model aircraft. Soon thereafter he reported to the 62nd Bomb Squadron (Heavy) at Barksdale AFB, Louisiana. It was while in that squadron (or after a lateral reassignment to the 596th Bomb Squadron (Heavy) also at Barksdale) that he flew the B-52G model and the tales in this volume occurred. He upgraded quickly to instructor pilot and served as a full time instructor in the wing's "Training Flight" before moving to the wing's Standardization/Evaluation section.

While at Barksdale he was sent to Carswell AFB, Texas for cross-training into the B-52 D-model and served a short TDY tour to U-Tapao Royal Thai Air Field, Thailand. There his crew was quickly tapped for cell lead duties and they sat "conventional alert" in the days leading up to the withdrawal from Vietnam in April of 1975.

In 1978 he was assigned to Headquarters SAC in Omaha Nebraska where he "flew a desk" every day as a strategic mission analysis and occasionally piloted T-39s with the local Military Airlift Command's detachment. Following the headquarters tour and ACSC he was assigned to Dyess AFB, Texas as a B-52H pilot. Just as he was beginning to requalify as an instructor his flying career was cut short by a silly wiggle during a routine EKG for an annual flight physical. Soon he transitioned to other non-flying jobs including Command and Control, and Logistics.

In 1985 he returned to Headquarters SAC as the division chief of the Logistics Control Division in SAC's underground command post. Reassigned in 1988 to Thule, AB Greenland he was Deputy Base Commander, Inspector General, and Director of Logistics. In 1989 he was assigned as Deputy Director, Combat Support for the 28th Air Division at Tinker AFB, Oklahoma and served in 1990 as Deputy Commander, Resources for its 552nd Airborne Warning and Control Systems (AWACS) Wing, also at Tinker.

He retired in January of 1991 and quickly returned to Thule, Greenland to rejoin a Danish surgical nurse who had "stolen his heart". He now lives in a little old house out in the peaceful Danish countryside - and is happy that none of his "war stories" took place while actually at war - he didn't log a single minute of combat flight time in 22 years of active service. More than you might care to know can be found on his personal web site: www.texan.dk

Kenneth B. Sampson

Captain, USAF (Ret.) I studied Aircraft and Powerplant mechanics in high school and became a licensed A&P mechanic in 1957. I was an enlisted jet engine mechanic in USAF for two years and an enlisted USAF Cadet Candidate at the West Point Preparatory School for one year. I was a dual status A/2C and Air Force Cadet at the USAF Academy for four years graduating in 1964.

I tried pilot training but was not physically coordinated enough to be a pilot, so I became a B-52D Navigator and Radar Navigator. I flew a total of 363 B-52D combat missions. I flew 343 B-52D combat missions bombing South Vietnam, Laos, and Cambodia. I flew 20 combat missions over North Vietnam. I flew 312 combat missions as a navigator and 51 combat missions as a radar navigator.

I flew zero missions in Linebacker II, because I was at Carswell AFB for upgrade training when Linebacker II hit. As a USAF B-52 navigator / bombardier, I exercised my responsibility to make an effort to win the war and free the prisoners (of which 10 were my USAFA classmates) by aiming and dropping over 35,000 bombs, with the intention of killing as many of the enemy as I could with each bomb.

I was awarded one DFC, 18 Air Medals and a 300-mission patch. I flew 2, 205.7 combat flying hours and 3945.3 total flying hours in the B-52D in seven years. I was stationed at Amarillo, Homestead, March, and Dyess. I had eight Arc Light tours, one year on Guam, and two years at U-Tapao and Okinawa. I left U-Tapao in November 1973 for five months at Sheppard AFB hospital and was medically retired as a Captain for a nervous breakdown.

I have been married to my Thai wife for 33 years. We have between us five children, six grandchildren, and one great grand child.

Kenneth R. "Ken" Schmidt

Major, USAF (Ret.) Was born in Shattuck, Oklahoma, grew up in Woodward, Oklahoma, and earned his B.S. degree in Business Administration from Southwestern (Oklahoma) State College in 1970. He entered the USAF through Air Force Officers Training School in April, 1974. He was then assigned to Mather AFB, California, for Undergraduate Navigator Training (UNT) where he received training in the T-29 and T-43 aircraft. Upon graduation from UNT and earning his navigator wings in December, 1974, Ken continued his training at Mather completing the ASQ-48 Navigator/Bombardier Training (NBT) course in April, 1975, and was assigned to the 9th Bomb Squadron, 7th Bomb Wing, at Carswell AFB, Texas. He completed Combat Crew Training School (CCTS) and was assigned to a combat crew. At Carswell, he served as navigator, instructor navigator, and radar navigator in the B-52D and left Carswell in 1980. His next assignment was to the 23rd Bomb Squadron, 5th Bomb Wing at Minot AFB, North Dakota, flying in the B-52H. While at Minot, he served on a combat crew as a radar navigator and instructor. He was later assigned to the 5th BW as a Bomber Scheduler.

In 1983, Ken was transferred to Castle AFB, California, as an Academic Instructor at the 4018th CCTS and served in that position for two years. His flying at Castle was in the B-52G. He was then assigned as the Chief, Air Weapons Branch, and remained in that position until being transferred in 1987. His next assignment was to Offutt AFB, Nebraska, and he was assigned to the 2nd Airborne Command and Control Squadron (2nd ACCS) as an Operations Plans Officer flying onboard the EC-135C "Looking Glass". In 1989, Ken was assigned to HQ/SAC DOO and held various staff jobs in the DO community until SAC closed up shop in 1992. He was then reassigned to the 55th Strategic Recon Wing and worked in the Wing Inspector's Office until his retirement in 1993. He accumulated 3,651 flying hours: 155 hours in T-29/T-43 trainers; 2,681 in the B-52D, G, and H; and 811 hours (100 operational missions) in the EC-135C.

While in the Air Force, Ken completed his Master's Degree from Texas Christian University. He also completed Squadron Officers' School (by correspondence and in residence), Air Command and Staff College, and Air War College.

Ken remained in Papillion, Nebraska, after his retirement and is currently an assistant director in the Financial Aid Office at the University of Nebraska at Omaha.

Ken was happily married to Susan (Datin) for 34 years and has two daughters, Stephanie and Staci. Unfortunately, Susan passed away on 1 July, 2004, after a five-year battle with breast cancer.

I would like to dedicate my portion of the book to the memory of Susan Schmidt, my wife, lover, and best friend; and also to all the other

SAC wives who put up with our career choice. Without these loving and caring wives, we could not have done our jobs.

George Schryer

Senior Master Sergeant, USAF (Ret.) started flying in early 1966 as an A1C and was stationed at Seymour Johnson AFB, North Carolina. Prior to that I was a B-47 Bomb/Nav electronics tech at Lincoln AFB, Nebraska, from 1962 to 1966. Prior to that I spent five years in the Navy on board a destroyer as a Gunners Mate.

My first SEA tour was from October, 1969 to February, 1970. While at Seymour Johnson I was an Instructor and then Stan/Eval. In April of 1971 I went PCS to Loring AFB, Maine and started my second SEA tour in Late November 1972. I had been assigned to this Stan/Eval crew for almost a year.

I am still married to my wife Gail of 37 years and we have two daughters and five grandchildren. I retired from the AF in 1980 after serving a tour as Staff Instructor at the USAF Survival School at Fairchild AFB.

Pete Seberger

Major, USAF (Ret.) Graduated from the University of Nebraska in June of 1962 with a BSME and a commission as a 2nd Lieutenant from AFROTC. He entered pilot training at Vance AFB that November and graduated in December of 1963 with class 64-D. He was assigned to SAC at Barksdale AFB to fly B-52Fs and after CCTS, survival, and nuclear weapons schools he began crew life in June of 1964. His unit was sent TDY to Guam in early 1965 and was transferred to Carswell AFB in May-June of that year. The combined wing then flew most of the early Arc Light missions until December of 1965 and in mid 1966 the D-model fleet took over those duties. In April of 1968 the Carswell unit was reduced to one bomb squadron and about half the personnel were dispersed to other units in SAC. Pete went to Grand Forks AFB, upgraded to A/C and flew two RTU tours before upgrading to IP. In 1973 he was assigned to Castle AFB as a flight line instructor for most of that tour. In 1977 he was assigned to 13th Air Force at Clark AB where he filled the remote tour square in several different jobs. He was reassigned to Ellsworth AFB (SAC never lets go) as a Crewdog, requalified as an A/C and IP simultaneously, and in June of 1979 was selected to command the

Physiological Training Unit at Ellsworth, while remaining as an attached instructor to the two squadron bomb wing. (Yeah! No alert but lots of flying!) In early1983 the unit was reduced to one bomb squadron so he went off flying status. He retired in December, having logged 6,600 hours flight time in the B-52, some of it every year between 1964 and 1983.

After his military service he worked for a small flight operation in Rapid City and earned his ATP and civilian instructor ratings, flew a year or so for a small airline, and finished his flying career as a Flight Safety instructor in the Beechcraft King Air series at the factory school in Wichita. His logbook is just shy of 9,000 total flight hours, over 4,000 as instructor.

Theron "Buddy" Sims

Lt Colonel, USAF (Ret.) Served in the USMCR 101st Rifle Company, Topeka, Kansas, as an infantryman, 1962-1965, and during the Cuban Missile Crisis at Camp Lejeune, North Carolina. He earned a BBA in Business in 1967 from Washburn University where he also earned a commission as a 2nd Lieutenant through the two-year AFROTC program. He attended Undergraduate Pilot Training at Enid AFB, Oklahoma. After pilot training, he completed a tour in Southeast Asia as a night Forward Air Controller (FAC) in the 0-2A aircraft completing 250 combat missions in 1968-1969.

After CCTS training in the B-52F model in 1969 at Castle AFB, California, he was assigned to the 17th Bomb Wing/34th Bomb Squadron at Wright Patterson AFB, Ohio flying the B-52H, 1969 – 1974 as a CP, P, and EP/IP. He deployed for six months to Guam/U-Tapao from May – October of 1972 as part of "Operation Arc Light" and was assigned to the 307th SW as an aircraft commander. He completed 100 B-52D combat missions over South and North Vietnam during the Linebacker I timeframe. Additional assignments in bombers included Barksdale AFB in 1st CEVG, 1974-1976, as instructor and evaluator pilot; Grand Forks AFB, 34 Bomb Squadron, 1979-1983, as pilot, tactics officer, standardization pilot, and Wing Director of Training; and Barksdale AFB in 8th AF, 2002-2003, as a B-52 staff pilot in the Combined Air Operations Center – Training (COAC-T).

During his career, he logged over 4,500 hours in the B-52F/D/G/H aircraft and over 7,000 hours total flight time in military and commercial aircraft including the B-727 and BAE-3100. His 20 decorations include the Distinguished Flying Cross, the Air Medal with 16 oak leaf clusters, Joint Service Commendation Medal, Global War on Terrorism Expeditionary Medal, Meritorious Service Medal, Air Force Commendation Medal, Meritorious Unit Award, Presidential Unit Citation, the South Vietnam Service Medal, the South Vietnam Cross of Gallantry, and South Vietnam Medal of Honor 1st Class. He currently lives in Edwards, CO with his wife Bonnie and is the Commander of the Veterans of Foreign Wars Post 10721, a member of the Edwards Rotary Club, and Vice President of the Freedom Park Memorial Committee.

Robert B. Stewart Jr.

Lt Colonel, USAF (Ret.). AB and MA degrees in Education Admin & Supervision, University of Kentucky, 1946-1951. After teaching for one year, he was facing a draft call and enlisted USAF for four years, in August of 1951. After completing basic training he was assigned as a technical instructor at Lowry AFB, Colorado, in Officer Personnel and Career Guidance School. Discharged in August of 1952 to accept reserve commission with direct appointment and no break in service. Completed navigation flight training at Ellington AFB, Texas, in 1953. Completed ECM flight training at Keesler AFB, Mississippi, in 1953. Assigned to 67 Tactical Recon Wing, K-14, Kimpo AB, Korea, in January of 1954 and 1st Marine Air Wing. Assigned to 548th Recon Tech Sq, Tachikawa, Japan, July of 1954. Assigned to Hq FEAF, Tokyo, Japan, July of 1954 with RC-135 mission planning and ECM intelligence collection duties. Assigned to Loring AFB, Maine, as Wing OJT Coordinator in 1956. He became a victim of rheumatoid arthritis and was hospitalized at Chelsea Navy Hospital and Walter Reed Hospital for six months. At that time he elected to remain on active duty. He volunteered for assignment to Keesler Tech Training Center, Mississippi in 1957 as ECM classroom and flight instructor, intelligence librarian and as a briefer on a traveling USAF ECM familiarization team. One of two officers selected from 15,000 assigned to ATC to receive a regular commission. Assigned Glasgow AFB, Montana, with B-52 crew duties as IEW and standardization evaluator from 1957-1962. This included a six-month TDY Arc Light assignment to Guam. Then he was surprised with an assignment to Hq 15AF, March AFB, California, as staff officer for EW training and operations, 1967-1971. Assigned to Dyess AFB, Texas, as 96 Bomb Wing, Chief of Penetration Aids, 1971-1975. During this time he was TDY to U-Tapao AB, Thailand, as B-52 mission briefer. Upon returning he augmented the crew training staff at Carswell AFB.

Retired in September of 1975 with 24 years of continuous service. Logged 5,654 flying hours, mostly in B-52 aircraft, with 630 hours of combat time. Logged ECM time in B-26, C-45, C-47, AD-4N, KC-97, C-54D, T-39 and B-52D. Citations include The Meritorious Service Medal, Commendation Ribbons, Air Medal with one Oak Leaf Custer, National Defense Service Medal, Korean Service Medal, United Nations Service Medal, Air Force Outstanding Unit Citation Award with one OLC, Air Force Longevity Service Award with 4 OLC, Good Conduct Medal, and Vietnam Service Medal. Currently resides with his wife, Jean, in Versailles, Kentucky, near Lexington.

George R. Thatcher, Ed.D

Major, USAF (Ret.) George Thatcher joined the Air Force in 1951, soon after graduating from his hometown high school in West Orange, New Jersey. He spent five years as an administrative specialist and training NCO, then attended and graduated from Air Force OCS Class 1957-B. After two years as an Air Defense Weapons Director, he was accepted into pilot training and graduated in Class 1960-H.

George chose to fly the "heavies" and was assigned as a B-52 copilot at Turner AFB, Georgia, where he spent over six years, upgrading to Aircraft Commander along the way. He flew many airborne alert missions, especially during the Cuban missile crisis, and had eight months of combat flying out of Anderson AFB, Guam. When Turner AFB closed in 1967 he was assigned as Aircraft Commander and Instructor Pilot to Plattsburgh AFB, New York.

In 1969 he volunteered for another tour of duty in Southeast Asia, this time flying the EB-66 "Destroyer" at Takhli RTAFB, Thailand. Following that experience, he spent his final two years of active duty as a Command and Control Officer at Blytheville AFB, Arkansas.

Since retiring from the Air Force in 1972, George has been a real estate broker, mortgage loan officer, middle/high school Spanish teacher, prison educator, and university professor of Education. He is retired from Texas Tech University and presently serves as Adjunct Professor of Workforce Education for Southern Illinois University's off-campus degree program. His major passion is writing/collecting aviation poetry and short stories, and he engages in a variety of fitness activities. He gives up the game of golf frequently.

George is married to Jean Schwisow-Thatcher, his Beauteous Wife and Princess for Life, whom he credits with teaching him everything he knows about the Education profession.

Tommie N. Thompson

Lt Colonel, USAF (Ret.) He qualified for the AF Aviation Cadet Program. After successfully completing pilot training, he was commissioned 2nd Lieutenant on April 29, 1960. He was assigned copilot duties in B-47s at Little Rock AFB, Arkansas. Following that, he upgraded and flew B-52s at Glasgow AFB, Montana. After a tour at Texas A&M University where he completed undergraduate and graduate degrees in civil engineering, he was assigned to Bergstom AFB, Texas for transition training in the RF-4C. His combat duty station was Udorn AB, Thailand where he flew 118 combat missions, 23 over North Vietnam. He was next assigned at 7th AFHQ. Upon completion of Captain Thompson's Southeast Asia tour, he was again assigned to Bergstrom AFB as an RF-4C pilot. His four-year tour there included upgrade to instructor pilot, squadron deployment to Avaino and several temporary assignments back to Udorn, Thailand.

Major Thompson was then assigned to Griffiss AFB, New York as Director of Operations and Maintenance for the base civil engineering squadron. Following that, he spent a year as the commander of Wake Island. This assignment proved particularly interesting in that it was the year the Vietnam refugees were transported to the US. Over 15,000 came through Wake Island. Following that, a year was spent at the AF Air Ground Operations School at Eglin AFB. Next it was "North to Alaska" where Lt Col Thompson served as the Director of Operations and Maintenance, Director of Engineering and Construction, and Director of Programs for the Alaska Air Command.

Lt Col Thompson retired on April 1, 1981. His awards include two DFCs, six Air Medals, Meritorious Service Medal. Bronze Star and three AF Commendation Medals. Tommie and his wife, Dawn, live in Salado, Texas.

Harry Tolmich

Captain, USAF (Ret.) Enlisted in the USAF 1954. Assigned Fairchild AFB, Gunnner on the B-36 Peacemaker. Flew the longest B-36 mission with Lt Col Harvey Downs in 1957, 46 hours and 10 minutes non-refueled. Transistioned to B-52 May of 1957. Tail Gunner for 10 years, TDY to Arc Light 1966 and 1968 from Glasgow AFB, Montana. Attended Park University on the Bootstrap program. Commissioned in December of 1970. Assignment was Maintenance Officer for the Boeing T-43 (737) aircraft at Mather AFB, California. Upon retirement as a Captain I was hired by the Boeing Company as a Maintenance Operations Manager to oversee the operation of the Boeing T-43 aircraft at Mather AFB. Sweet!! I retired from Boeing with 23 years service in 1995. My wife Lois and I make our home in Las Vegas, Nevada.

Tommy Towery

Major, USAF (Ret.) earned a B.S. degree in Journalism in 1968 from Memphis State University where he also earned a commission as a 2nd Lieutenant through the AFROTC program. Attended Navigator Training and Electronic Warfare Officer Training at Mather AFB, California. Following B-52 CCTS training in B-52F models at Castle AFB, California he was assigned to the 20th Bomb Squadron at Carswell AFB, Texas flying B-52C and B-52D models. He served as a 7th Bomb Wing Combat Intelligence briefer and a Penetration Aids staff officer while grounded from flying status for kidney stones. He was deployed for six months to Guam as part of "Operation Bullet Shot" and was assigned to 8th Air Force Bomber Operations as an Arc Light mission planner. Three month after returning from his first deployment he was sent on a second six-month TDY to Guam where he worked as an Arc Light planner in the 43rd Bomb Wing Bomber Operations. It was during this assignment that he flew on B-52 combat missions as a staff officer and helped plan Linebacker II missions. Upon return to full flight duty status he was assigned to a B-52D crew at Carswell AFB, Texas. Shortly thereafter he deployed to Guam and Thailand with his first crew and later progressed from crew duty to instructor to standboard duty on Crew S-1. In 1976 he and his crew received the Mathis Trophy, awarded to the top bomber unit based on combined results in bombing and navigation in the SAC Bomb-Nav Competition.

Following his B-52 assignment he was transferred to the RC-135 program at Offutt AFB, Nebraska and flew operation reconnaissance missions from forward operating bases. In 1983 he was assigned to a four-year tour as an Electronic Warfare Intelligence officer and briefer at RAF Mildenhall, UK. Upon his return to the states he spent his last year of active duty as an Electronic Warfare Intelligence officer with the 55th Strategic Reconnaissance Wing at Offutt AFB, Nebraska.

During his career he logged over 1,600 hours in B-52s and over 5,000 hours total flight time. His decorations include the Meritorious Service Medal with oak leaf cluster, the Air Medal with eight oak leaf clusters, the Armed Forces Expeditionary Service Medal, the Republic of Vietnam Campaign Medal and the South Vietnam Cross of Gallantry with Palm. He currently is employed as a computer specialist at the University of Memphis and lives in Memphis, Tennessee, with his red-headed wife Sue who graciously allows him to devote many hours to his book writing hobby. Tommy has written two non-military books, *"A Million Tomorrows –Memories of the Class of '64"* and *"While Our Hearts Were Young."* Before contributing to and editing this book, he was a writer and editor of "*We Were Crewdogs – The B-52 Collection*" and *"We Were Crewdogs II – More B-52 Crewdog Tales"* and is a member of The Military Writers Society of America.

Jimmy Turk

Major, USAF, (Ret.) Earned an A.A. in Pre-Engineering in 1959 and entered the USAF Aviation Cadet program where he earned his commission as a 2nd lieutenant and was awarded his Navigator rating. .Completed three Bomb/Nav training courses at Mather AFB, California. After CCTS training in B-52B/D/F models at Castle AFB, California and Land Survival Training at Stead AB, Nevada was assigned to the 449th Bomb Squadron, 4239th Strategic Wing (Heavy) at Kinchloe AFB, Michigan (an ADC Base). When he arrived the wing had only one loaner aircraft 60-0036 from Wurtsmith AFB, Michigan. He helped bring two new B-52Hs from the Boeing Wichita facility to the wing. During the Cuban Missile Crises he was an instructor navigator on 10 24-hour plus Airborne Alert Flights and later flew 10 more Airborne Alert flights as an emergency back-up communications link for the Thule AFB Ballistic Missile Early Warning (BMEWS) site. The wing was renamed the 449th Bomb Wing, 93rd Bomb Squadron. After representing the win in the 1965 World Series Bombing Competition at Fairchild AFB, Washington he was selected to be a part of the 80 Nav/RNs transferred to AFCS via Staff Officers Communications Course at Kessler AFB, Mississippi. Upon completion he was assigned to the Far East Communications Region Headquarters at Fuchu AB, Japan. Later he was assigned to EC-47 crew training at England AFB, Louisiana and the thru PACAF Jungle Survival School conducted at Clark AFB, PI. Upon completing this training he was inducted to the 361st Tactical Electronic Warfare Squadron (TEWS), 4560th TAC Recon Wing initially at Nha Trang AB, RVN which was moved to Phu Cat AB, RVN.

After completing his in-country tour he was re-assigned to Castle AFB, California for B-52 CCTS – this time as a Radar Navigator. He became an integral part of the34th Bomb Squadron, 17th Bomb Wing (Heavy) where he took part on one Arc Light tour and was promoted to the #2 Standardization/Evaluation Radar Navigator. He was later transferred to Air Force Logistics Command, and attended Computer Programming course at Sheppard AFB, Texas. Was transferred to Karamursel CDI, Turkey as the senior site programmer for a year and returned to AFLC for his last two years prior to retirement in July of 1979. He has over 5,200 hours flight time with over 3,500 in B-52s and over 1,000 hours combat time in EC-47s. His decorations include the Distinguished Flying Cross with one oak leaf cluster, the Air Medal with eight oak leaf clusters, Air Force Commendation Medal with one oak leaf cluster, Combat Readiness medal with two oak leaf clusters, Presidential Unit Award, AF Outstanding Unit Award with "V" device and two Oak Leaf clusters, Republic of Vietnam Campaign medal, and the Vietnam Service Medal with 7-BSS. He re-wrote and updated the Ballistic Training Manual T.O. 1-B-52D-25-1. He currently lives in Huber Heights, Ohio with his wife Alta.

Rich Vande Vorde

Major, USAF (Ret.) was commissioned a 2nd Lieutenant in 1971 through the AFROTC Program at University of Minnesota at St. Paul, Minnesota, where he received his MS in engineering. After graduation from Undergraduate Navigator Training and Navigator Bombardier Training (NBT) at Mather AFB, California, he was assigned to B-52F Combat Crew Training at Castle AFB (for Q-38 training). First assignment was McCoy AFB, Orlando, Florida, which promptly closed and then reassigned to B-52Ds at Carswell AFB, Texas (for Q-48 training). He flew as a B-52D Navigator and then upgraded to Radar Navigator at Carswell until his 1978 assignment to Mather AFB, California as an NBT instructor. In 1982 he was assigned again to B-52Gs at Anderson AFB, Guam, where he served on a crew, then as a Wing scheduler. In 1984 he was assigned to K.I. Sawyer AFB, Michigan where he flew B-52Hs and was a scheduler until 1987, when he was assigned to McConnell AFB, Kansas as a scheduler for the newly arrived B-1B bombers. He retired four years later in 1991. He and his wife, Cheri, are enjoying retired life now in Paradise, Texas, as well as visiting their son Tim, daughter Traci and grandchildren. They also do a lot of traveling, through the military (Space-A) program and cruising.

Marvin Zane Walker

Major, USAF, (Ret.) Graduated from the aviation cadet program, Class 59-07N, with 2nd Lieutenant officer's commission and navigator rating in spring of 1959. Attended Advanced Navigation, Reconnaissance, Bombardment class at Mather AFB, and completed it in fall, 1959. Assigned to Strategic Air Command (SAC) 96th Bomb Wing B-47's at Dyess AFB, earning spot promotions to 1st Lieutenant after becoming combat ready and later was awarded a spot promotion to Captain. Reassigned to 91st Bomb Wing and B-52D's at Glasgow AFB, Montana, in the summer of 1963, and in 1966, his first Vietnam missions were flown out of Guam. Assigned in 1967 to Air Force Institute of Technology mathematics/computer science program, obtaining B.S. from Colorado State University in 1969. Then served a year at Da Nang AB in Vietnam, flying over 200 combat missions in Laos and Vienam in AC-119 K (Stinger) gunships in the 18th Strategic Operations Squadron, where he was awarded the Distinguished Flying Cross and several Air Medals. Reassigned to 96th BW B-52D's at Dyess AFB, Abilene, Texas, in April, 1971, there obtaining his instructor navigator rating and master navigator rating. Flew over 100 Vietnam (Arc Light) missions out of Andersen AFB, Guam and U-Tapao AB, Thailand. After the Vietnam war ended, he became target study officer in the 96th Bomb Wing Bombing/Navigation Section. Final assignment in 1975 was to SAC Headquarters at Offutt AFB as a computer programmer/analyst to create automated B-52 war plans. Retired Sept. 1, 1979 with 5,000 hours flight time, including 2,000 combat hours. He moved to Albuquerque, New Mexico, to take a job as a computer programmer/analyst with Public Service Company of New Mexico, where he worked for 18 ½ years. He and his wife, Phyllis, still live there. He met Phyllis at Mather AFB at the beginning of his military career. They have one son, Robert, who now lives with his family in Mesquite, Texas.

Gates Whitaker

Former Captain, USAF. Served in the Air Force from 24 Sep 69 to 28 Jun 74. Flew 130+ combat missions as copilot on Dyess E-15. Also sat alert at Dyess and Bergstrom. Came into the Air Force through ROTC at Union College, Schenectady, New York. I had never been off the East Coast until summer camp at Gunter AFB in Montgomery, Alabama, and UPT at Laughlin AFB in Del Rio, Texas. Talk about culture shock - going out to Del Rio, the cactus got bigger than the trees!

Met and married a Texas native Mary McDonald while stationed at Dyess. That was the best thing that happened to me from 1971 to the present . She lived in Guam and Thailand for a while, even going to Cambodia to renew her visa. We moved to Dallas after my discharge and raised three boys. Since then, I have had an "under the radar" career in the packaging and food manufacturing business.

About our crew - we flew six missions during Linebacker II. I don't know of anyone else who had that schedule. Of the nine people I flew with in three years on the crew - I think seven were career Air Force, some after detours in and out of civilian life. They were all great as individuals, and on the crew it got to where we could finish each other's sentences. My time in the Air Force was unforgettable, and I'm glad to be able to add a few stories about it.

Bruce Woody

Former Captain, USAF. Entered the Air Force in June, 1970. He was commissioned through the Air Force ROTC program at the University of Nebraska-Lincoln, where he was a Distinguished Military Graduate and Commander of his AFROTC Cadet Corps.

After spending a year and a half attending Undergraduate Navigator Training and Navigator/Bombardier Training at Mather AFB, he attended B-52 transition school at Castle AFB. He was then assigned to Dyess AFB as a Squadron Navigator, arriving at Dyess in February of 1972.

Bruce participated in 160 Arc Light combat missions, including three missions in Operation Linebacker II. He was awarded the DFC, nine Air Medals, Presidential Unit Citation, Cambodian Expeditionary Medal, and the Republic of Vietnam Cross of Gallantry with Palm. He participated in the last B-52 combat cell of the war, over Cambodia, shortly before midnight local time on 15 August 1973.

Returning to Dyess in 1973, he was sent back to Andersen AFB with his crew to establish a permanent Standboard Office for the peacetime Wing.

In 1974, Bruce went back to Dyess for good, where he was named as navigator on a Select Crew, and upgraded to Radar Navigator in February, 1975. He left the Air Force in 1976 to attend law school. He has been an attorney in Texas for 27 years, 12 of which were served as a judge.

Bruce and his wife Jan have been married for 30 years.

John York

Lt Colonel, USAF (Ret.) Was a distinguished graduate of Officer Training School and attended Undergraduate Pilot Training at Laredo AFB, Texas. He flew the highly subsonic O-1E in Vietnam as a Forward Air Controller (FAC). He flew the B-52G at Barksdale AFB, Louisiana, Wurtsmith AFB, Michigan, and Barksdale AFB, Louisiana. He also flew the T-39 and KC-10. He was a Command Post "Weenie" and Accounting and Finance Officer. Following retirement from the USAF he flew at Federal Express until he got so old the FAA wouldn't let him fly anymore. He now enjoys farming, ranching, hunting, and fishing in Oologah, Oklahoma. He and his lovely wife Ruthie have raised two fine sons. He can be reached at OldChkPilot@Yahoo.com.

Personal Notes